PHYTOSOCIOLOGY OF KASHMIR HIMALAYAS

Edited by:

Aasif Ali Gatoo (Ph D)

Assistant Professor
Division of Natural Recourse Management
Faculty of Forestry (SKUAST Kashmir)
Benhama Ganderbal Kashmir (J&K)

United Book Prints

United Book Prints

Reprinted in India by:

UNITED BOOK PRINTS
4806/24, Ansari Road, Daryaganj,
New Delhi 110 002, INDIA
E-mail: info@scientificpub.com
Website: www.scientificpubonline.com

Print 2025

ISBN: 978-93-95367-09-7 (H/B)
eISBN: 978-93-95367-11-0 (E/B)

Printed in India

Dr. Aasif Ali Gatoo
Assistant Professor
Division of Natural Recourse Management
Faculty of Forestry (SKUAST Kashmir)

PREFACE

India is one of the 17 mega-diverse countries because of the richness of biodiversity. Around 45500 plant species are present in our country that means it a rich bowl of floral diversity. The Himalayas is known as one of the world's biodiversity hotspots because of the richness in both flora and fauna. The Kashmir Valley, also known as the Vale of Kashmir, is an intermountain valley concentrated in the Kashmir Division of the Indian-union territory of Jammu and Kashmir. The valley is bounded on the southwest by the Pir Panjal Ranges and on the northeast by the main Himalayas range. This area holds plant communities of various origins sand characteristics. To study more about plant communities the science of phytosociology is obligatory. Phytosociology is the discipline which concerns itself with the study of vegetation as such, with its floristic composition, structure, development and distribution. The study of species diversity is helpful in understanding community composition, structure, change and development. It is the subset of vegetation science that deals with extant plant communities and puts particular emphasis on their classification.

This book is an approach towards understanding the phytosociology of Kashmir Himalayas. It highlights the interrelationship between various plant communities based on their phytosociology.

The Editor is thankful to Dr. S. A. Gangoo, Dean Faculty of Forestry; Dr. Akhlaq Amin Wani, Professor and Head Division of NRM and his fellow colleagues Prof. M A Islam, Dr. Shah Murtaza, Dr. Amir Bhat, Dr Khurshid Ahmad Sofi, Dr M Maqbool Rather and Dr Megna Rashid for their immense support and guidance throughout the journey. Also, the author thanks his family and students for their consistent inspiration and encouragement.

Dr Aasif Ali Gatoo

Prof. S. A. Gangoo
Dean, Faculty of Forestry
Sher-e-Kashmir
University of Agricultural Sciences and
Technology of Kashmir

MESSAGE

The hidden treasure of biodiversity that resides in the Kashmir Himalayas is of great value and Importance. The interrelationships of plant unravel the many mysteries of nature in general and particularly biodiversity. More we know, more benefited we will be. This book is an excellent approach to study the phytosociology of Kashmir Himalayas. It helps in understanding the plant communities and their occurrences to enable us to greatly be benefitted by the overall vegetation. It is important tool for getting an overview of the species richness of the country for ecological stability.

Dr. Aasif Ali Gatoo has donc a commendable job for unraveling the nature and it's mysteries.

I extend my compliments and best wishes to him for his amazing book on "Phytosociology of Kashmir Himalayas". This is an achievement for the Faculty of Forestry, Sher-e-Kashmir University of Agricultural Sciences and Technology of Kashmir to have this piece of work as a result of his hard work and consistency.

Dean, FoF

Dr. Akhlaq Amin Wani
Professor and Head
Division of Natural Recourse Management
Faculty of Forestry (SKUAST Kashmir)

FOREWORD

The book "Phytosociology of Kashmir Himalayas" by Dr. Aasif Ali Gatoo highlights the various aspects of plant communities of Kashmir Himalayas. The ecological importance of plants and their interrelationship forms the basis of all other sciences. Phytosociology as discipline concerns itself with the study of vegetation as such, with its floristic composition, structure, development and distribution. The science of physiology helps in understanding the key aspects of plant communities which ultimately is important for understanding the role of plant communities in ecological stability.

The mountains of Himalayas being one of the biodiversity hotspot in the world have a lot to offer. Kashmir Himalayas are rich in all sorts of biodiversity. This book is a elaborated source of knowledge about the plant communities in Himalayan Kashmir. It is such a pleasure for me as HOD of Natural Resource Management to have such a fine piece of work being done by one of the finest scientist, Dr. Aasif Ali Gatoo. I extend my compliment to the scientist for this absolutely needed piece of research.

Dr. Akhlaq Amin Wani

CONTENTS

Chapter 1

Current status of Agriculture and allied sectors in Kashmir their constraints and future possibilities

Oyais Ahmad Wagay[1], Jauhar Rafeeq[1], Mir Adil[2], Mohammad Kaif[3], and Tanveer Ahmad Rather[1]***

Division of Silviculture and Agroforestry, Faculty of Forestry SKUAST-Kashmir[1]
Division of Forest Biology and Tree Improvement, Faculty of Forestry SKUAST- Kashmir[2]
Division of Natural Resource Management, Faculty of Forestry SKUAST- Kashmir[3]

**Email: owaiswagay777@gmail.com*

Abstract

Agriculture being the backbone of State economy and important contributor to SGDP is losing its status as it is subject to different constraints. The current situation is not satisfactory in terms of food grains. The area under crops has shown the disturbing trends. The farmers are losing their interest in farming sector because it has become less remunerative as a result the existing population is not able to cover the gap between deficit and requirement which has already gone to 81%. The scope of further increase in the area under these crops is less as they are not found of profitability. For ensuring inclusive growth still agriculture needs to get priority in order to make adequate dent to poverty reduction and providing food security to the people. (Kaloo and Choure, 2021). Agriculture is the mainstay of the people as it provides employment directly or indirectly to about seventy per cent of the work force. (Maqbool and Jan, 2017) The increase in area, production and productivity of major food crops in Jammu & Kashmir State of India over the years but the rate of increase has been slow. Jammu region of the State has maximum cropping intensity (176.8%) followed by Kashmir (123%) and Ladakh (106%). The average land holding in the State is very low (0.66 ha) in comparison to National Average (1.68 ha). The bottleneck in mechanization are due to non-availability of improved equipment, small and scattered land holding, low investing capacity of the farmers and lack of awareness among farmers. Despite these, there is a great need and effort to increase the productivity and reduce the drudgery through mechanization inputs. (Dixit and Ali, 2014.) According to the Forest Survey of India's latest 2019 report, the forest cover in Jammu and Kashmir increased by 370 sq km in comparison to the forest cover in 2017, but the figure has been disputed even by the government's own experts.

Keywords: Cropping intensity, Deficit, Food grains, Jammu and Kashmir, mechanization.

Introduction

Jammu and Kashmir is a hilly state having a varied topography with a large diversity in terms of culture, social and economic practices of its different regions (Jammu, Kashmir and Ladakh). Despite the structural changes taking place both at national and state level, agriculture has been the top priority at both the levels as this sector plays a strategic role in the process of economic development of Jammu and Kashmir and India, and on an average 70% of the population are still getting livelihood and employability from this sector coupled with its allied sectors as this sector contributes around 27% of the state's income ((Kaloo and Choure, 2021).

Presently agriculture sector contributes 13.7% of the GDP at national level and at state level it is contributing around 21.09%. The distinct geographical outlook determines the cropping pattern and productivity of crops. Jammu and Kashmir is well known for its Paddy crop followed by maize, oilseeds pulses, vegetables fodder and wheat, where as in Jammu region the most eatable and produced crop is wheat which is followed by maize, paddy, pulses, oilseeds etc whereas barley is the major cereal crop followed by wheat in Ladakh. Jammu and Kashmir have also got the monopoly in terms of Saffron crop (famous all over the world due to its quality) which is being produced in the district Pulwama 15KM away from Srinagar. At the same time Basmati Rice is being produced in R.S Pora of Jammu division and Rajmesh of Bederwah, Kishtiwar constitute the economic significance of the state which are functioning under the various departmental interventions in quality improvement, productivity enhancement, marketing etc.

The growth rate in respect to agriculture and allied sectors for the 12th five year plan has been targeted to be around 2.99% and as per the advances estimates growth rate for the financial year 2014-15 has been projected to be around 3.84% but at the same time the growth rate in the sector, particularly in the crop sub sector is on a decline as the percentage of deficit is continue on a high, to meet the requirements of the growing population which has resulted in the import of these gaps from the central pool. (Kaloo and Choure, 2021).

The agriculture and its allied sectors provide about 36% of the net domestic product and engage about 80% of the population. Draught animals are the main source of farm power. Animal drawn implements such as Indigenous, Tawi, Hill and Soil stirring plough are employed for land preparation activities. Small and marginal farmers still use the manual tillage tools such as spades, shovels, sickle, hand forks while farmers in plain topography use tractors and implements for tillage and transportation purposes (Dixit *et al.*, 2006; Sharma, 2010). Farm work is generally labour intensive and demands a considerable amount of energy. In general, the level of mechanisation in the State is low. Although there is abundance of water received through precipitation (snowfall and rainfall) in the State except in Ladakh region, the major portion is utilised in the lower part of the State and in Pakistan. Only 41.7% of the net sown area is irrigated by different sources and the remaining area is rain fed. The major source of irrigation is through canals (93.0%). Similarly, the natural energy sources like sunshine, wind, vegetation, water flow and other biological wastes which are abundantly available in the State are not being harnessed properly, resulting in very low per capita energy availability, deforestation and poor health.

The net domestic produce increased from Rs. 1049.5 crores at 1980-81 current prices index to Rs. 24835.7 crores during 2006-07, while per capita income increased from Rs. 1776 to Rs. 22506 during the same period. There are 13.35 lakh land holdings in the State with a population of 101.44 lakh. Nearly 82% of the total geographical area is not available for cultivation whereas the remaining 18% area is being used for agriculture and other allied activities of which only 7.36% is the net sown area (Table 1). Agriculture sectors during the post-independence era have shown considerable transformation, whereas the net sown area has increased from 0.614 million ha to 0.736 million ha during the last more than 60 years. It is estimated that there is a net deficit of 0.35 million t food grain to ensure food and nutritional security of the State population.

Table 1: Land resource utilisation in Jammu and Kashmir (2006-2007)

Classification	Area (million ha)	% of Revenue Records Area	% of Geographical Area
Total geographical area (surveyor general of India)	22.2236	-	-
Area under foreign occupation	12.0849	-	-
Net geographical area	10.1387	-	100.00
Area on revenue records	2.4160	100.00	-
Not available for cultivation (surveyor general of India)	8.3099	-	81.96
Not available for cultivation (revenue records)	0.5820	24.09	5.74
Area under forest	0.6580	27.23	6.49
Uncultivated land including pasture	0.3450	14.28	3.34
Fallow land	0.0890	3.68	0.89
Gross cropped area	1.1264	46.6	9.17
Net sown area	0.7420	30.71	7.36

Table 2 has been deployed to see the current scenario of Jammu and Kashmir in terms of food grain requirement in relation to population growth, and deficit. As from the table it has been estimated that right from 1947 the growth rate of population has surpassed the food grain production which has resulted in the mounting gap of deficit to requirement.

Looking at the current scenario right from 2011-12, the gap between deficit and requirement has shown an increasing trend from 1980-81 but during the year 2013-14 the gap between the former has declined up to a large extent. The current situation of Jammu and Kashmir is worse as there is a huge and a mounting gap of 81.50% of deficit compared to the requirement of the current population. The gap has been recorded as the highest compared to last year, as from the data the land diversification from agriculture to other commercial crops has gone up to 10 times. The present scenario is not good as there is no alternate and the per capita consumption is on a decline and by the time it is going to increase which is going to compel the people to import and in the coming two to three years there will be 100% deficit in terms of food grains which will mount the pressure on the economy as there is a mismatch between demand and supply which is going to led into inflation.

Table 2: Food Grain Production in Jammu and Kashmir in relation to population Growth Production/Requirement (Quantity =000 Tonnes)

Year	Production	Population	Requirement	Deficit%
1950-51	206.30	1795304	307.00	32.00
1980-81	486.92	3269276	559.05	23.00
2011-2012	826.72	7426575	1260.0	34.00
2013-2014	957.92	7488918(P)	1271.05	24.63
2014-15	240.00	7638696 (P)	1296.47	81.50

Source: Government of Jammu and Kashmir, Agriculture department of Kashmir Division.

Maize crop is almost the monopoly of Jammu division is continuously on a decline as it has shown a continuous decline in the field of area, production and productivity, wheat being the most cultivable crop in Jammu region in terms of consumption and economic value is gradually on a high and is earning the status of foreign exchange earnings, and employment opportunities.

Prices of the food grains have been rising at a very fast rate which have compelled the farmers to start food grains for the market, in the same way they grow oilseeds and other commercial crops or one can say that the traditional classification between food crops and commercial crops is losing its significance. Among the food grains the largest increase in area has been recorded in wheat, but the increase in the area under rice has been quite modest. The increase in the area of commercial crops/coarse cereals have recorded only marginal increase indicating a positive shift from minor to major crops. Increase in the area under pulses and barley has been modest. The traditional crops that are oilseeds, barley, pulses have made an impressive increase in acreage much more than the food crops (with the exception of rice) as they have become remunerative under the impact of new technology. All this is happening only due to increase in the economic values and the influences of prices subject to income.

Brief scenario of agriculture sector

The predominant sector of the economy of Jammu and Kashmir, supports about 80% of its population. Rice is the chief crop of Kashmir zone, followed by maize, barley and wheat. Jammu region dominates both in maize and wheat production. In the Ladakh region, barley is the major cereal crop followed by wheat. The production of three important food crops, namely, rice, maize and wheat, contributes a major portion of the food grain in the State and accounts for 84% of the total cropped area; the balance 16% is shared by other cereals and pulses.

Nearly 75% of the country's temperate fruits, mainly apples, are grown in the State. Crop diversification towards less perishable horticultural crops like walnut, almond, pistachio, olive, kalazeera, saffron and pomegranate in the region is essential to earn foreign exchange and reduce pressure on the available resources for creating additional infrastructure facilities. The State is deficient in food grains production. Improved technologies have been developed which are superior to the traditional practices in terms of productivity and profitability but their adaptation is not comparable to the adoption level in Indo-Gangetic regions of India.

The State needs to shift its agriculture development strategy from food security mode to the value addition mode by growing high value fruits, vegetables and some cash crops. There is good scope for the production of high value, low volume crops like saffron, black zeera and other spices especially in Kashmir region.

Paddy is the main crop of Kashmir, followed by maize, oilseeds, pulses, vegetables, fodder and wheat. In Jammu region, wheat is the predominant crop followed by maize, paddy, pulses, oilseeds, fodder, vegetables and other crops while in Ladakh, barley is the major cereal crop followed by wheat.

State also has the honour of being amongst the world's few places where quality saffron is cultivated. Pampore Tehsil of District Pulwama, which is situated at a distance of 17 kilometres from Srinagar, is famous for its high grade saffron in the world. Saffron is also grown, though on a limited scale in Kishtwar district of Jammu region. Saffron crop is being introduced in temperate areas of other districts as well and significant success has been reported from some areas of Doda and Udhampur districts. Due to the implementation of Saffron Mission interventions it has been possible for the state to rejuvenate the old less productive Saffron zones and it has been experienced that the new corm suggested has benefited the farmers by way of increased production leading to increase in their earnings and has created an effect of attraction for undertaking the farm business management for development of saffron.

For enhancing agricultural productivity on a sustainable basis, improving Seed Replacement Rate (SRR) is an important component. The Department of Agriculture has registered a 33% SRR of rice in Kashmir valley and 26.32% in Jammu Division. SRR in respect of maize was observed at 24% in Kashmir and 32.10% in Jammu Division. SRR in respect of Fodder is anticipated at 32% and 35% during Rabi Season in Kashmir and Jammu respectively. An estimation made by the Directorate of Agriculture Jammu, food grain requirement for the State works out to 25 lac MT per annum as against the tentative production of about 20.65 Lac MT during 2013-14. Therefore, the State faces a deficiency of 4.35 Lac MT of food grains which is provided by the CAPD through the rationing system. During the year 2014-15 food production got reduced to 17.42 lac MT due to devastating floods in the State.

Table 3: Comparison of Crop Yield of J&K State with the largest producing Indian States and some other Countries for the year 2012 shows the agricultural backwardness of the state as under.

S. No	Crop	Yield of J&K State (Qtls/ha)	Highest Yield States in India	Yield (Qtls/ha)	World Highest Yield Countries	Yield (Qtls/ha)
1	Paddy/Rice	20.85	Punjab	39.52	China	66.61
2	Wheat	14.04	Punjab	50.17	UK	73.60
3	Maize	16.48	Tamil Nadu	53.72	USA	88.58
4	Oilseeds	8.24	Gujarat	17.23	UK	35.88

Horticulture

Horticulture J&K State is well known for its horticultural produce both in India and abroad. The state offers good scope for cultivation of horticultural crops, covering a variety of temperate fruits like apple, pear, peach, plum, apricot, almond, cherry and sub-tropical fruits like mango, guava, citrus litchi, phalsa and Berete. Besides, medicinal and aromatic plants, floriculture, mushroom, plantation crops and vegetables are cultivated in the state.

Apart from this, well known spices like saffron and black Zeera are also cultivated in some pockets of the state. As a result, there is a perceptible change in the concept of horticulture development in the state. There are around 7 lakh families comprising about 33 lakh people which are directly or indirectly associated with horticulture. Horticulture development is one of the thrust areas in agriculture and a number of programmes have been implemented in the past, resulting in the generation of higher incomes in the rural areas, thereby improving the quality of life in villages. Looking at the specie-wise details of area figures under fruit for the year 2013-14, it is observed that the largest area of 45.34% (161364 hectares) is occupied by apples out of total area of 355921 hectares under fruit. Walnut is the next major fruit occupying 26.86% in the overall area under fruit.

The state holds first position in the production of temperate fruits like apples. As per the latest information available for the year 2017-18 production of fruit was 22.53lakh metric tonnes which accounts for around 2.47% of total fresh fruit of 948.84 lakh tonnes for 2017-18(PE) produced in the country. Export of fruits outside the state during 2017-18 was 19.78 lakh tonnes (including 0.23 lakh tonnes of dry fruits) constituting 84% of total fruit production. In fact, export of fruit forms the base of the state's trade. As per figures for 2017-18, the Horticulture department has reported 330 thousand hectares of land 19 under fruit cultivation compared to 6373 thousand hectares in the country for 2017-18.

Brief scenario of fisheries sector

There are 30,453 fishermen and 10,000 workers are employed in the fisheries sector which contributes 1.92 lakh quintals of fish production in the State during the year 2006-2007. However, there has been a big gap between demand and supply of fish. The State has potential of producing over 40 million t of fish along the length of 27,781 km of rivers and streams in the State. There are 18 State-owned trout hatcheries and 22 fish farms. There is good scope for expanding employment opportunities through the development of this sector.

Fisheries sector occupies a very important place in the socio-economic development of the state. It has been recognized as a powerful income and employment source of cheap and nutritious food, besides being a source of livelihood for a large section of the economically backward population of the country. The total fishermen population in the State as per livestock census 2003 was around 31,000. It is presently estimated at 93000. The 27781 Km. length of rivers/streams facilitate farming of more than 40 million tonnes of fish. As against this, the State has only 0.07 lakh hectares under reservoir area. There is a big gap between the demand and supply of fish. Fish is a valuable element of the diet of the local people throughout the year. There is also a demand for fish from the defence personnel and tourists. There are 1248 lakes including water bodies and water is spread into 0.40 lac hectares of area which gives an indication of the potential for fisheries in the State.

Brief scenario of livestock sector

The State has a precious wealth of livestock in forms of cattle-buffalo, sheep, goats, poultry, etc. The cattle and poultry amongst all the livestock are considered the most important tools for the development of the rural economy. On an average there were 6.38 animals per household (livestock census-2003). The production of pashmina shawls and other animal products like carpets, shawls and blankets of Kashmir earn handsome foreign exchange for the nation. Livestock activity contributes about 11% to the State GDP. Therefore, the livestock industry in the State has vast scope for development rendering quick economic returns. In terms of livestock production, there is a gap between demand and supply.

Brief scenario of dairy sector

The milk production has increased from 3.69 lakh t in 1995-1996 to 6.66 lakh t in 2001-2002, but the demand for milk and milk products has increased at a faster rate. There is great scope for dairy development in the State. The State is lagging far behind the State of Haryana and Punjab due to non-replacement of low milk yielding local breeds of cattle by the introduction of some high milk yielding new breeds. The sector has great potential for creating considerable job opportunities.

Livestock Integrated Sample Survey (2011-12) revealed that the total livestock population of Jammu region had demonstrated an increase of 8.40% over the year 2009-10 and reached to 78.908 lakhs from 72.79 lakhs. Kashmir region also witnessed an increase of 4.63% as livestock population had reached 74.994 lakhs from 71.67 lakhs. However, in Ladakh, the livestock population slipped by 1.44% from 6.60 lakhs to 6.505 lakhs during the same period.

The latest livestock census (2007) reveals that the state has a contribution of about 2% in the total livestock of the country. Livestock per hundred of population in the state is 88 animals as against only 46 animals in the country. Moreover, average livestock available per household in the 21 states is 7 animals, as against only 3 at all-India level. Both these indicators reveal that the magnitude of livestock per 100 of population and per household available in the state is twice its magnitude available at the National level.

Milk production of the state for 2016-17 was at 23.76 lakh metric tonnes. Per capita availability of milk works out to 451 grams per day as compared to 355 grams for the country in 2016-17. Dairy is the mainstay of the livestock sector in the state. The number of poultry birds as per

19th livestock census (2012) was 82.73 lakh in the state as compared to 7292.09 lakh in India. Poultry development is one of the most resilient sectors in the country fast growing itself to changing bio-security, health and food safety needs. Egg production of India for the 22 Year 2016-17 has been 881369.56 lakhs with per capita availability of 69 eggs per annum. This indicator for J&K is 18.38 eggs per capita per annum with egg production of 2305.29 lakhs for 2016-17.

Floriculture

Floriculture has a vast scope and potential in the valley, which is evident from the fact that during the year 1996, an area of 80 Ha was under flower cultivation in J&K, which has currently expanded to the level of 350 Ha, with an annual turnover of about Rs.1350 lacs. Further, more than 1500 youth are directly engaged under the commercial floriculture sector in Kashmir Division. Moreover, an area of about 9.297 Hac. has been covered under protected cultivation, in the private sector by raising Tubular Structure Poly Houses; High Tech Poly Houses, Shade Net Houses, etc. Development of commercial floriculture has been assigned a top priority by the Government, under Technology Mission and RKVY, as it holds a tremendous potential for employment generation.

Forests

Forests Jammu and Kashmir has a total forest cover of 20230 Sq km (20.23 lakh hectares) accounting for 19.95% of the total Geographical area of 101387 Sq km on this side of the Line of control. All India figures indicate a percentage of 24.47% geographical area under forests. Out of the total forest area of 20230 Sq km, the area under reserved 9 forests is 2551 Sq km, protected forest is 17463 Sq km and the rest 36 Sq km is unclassified.

Out of the total forest area, 12066 Sq km are in Jammu Region, followed by Kashmir Region with 8128 Sq km and Ladakh Region with only 36 Sq km. Kashmir Region has 40.17%, Jammu Region has 59.64%, while Ladakh Region has 0.17% area under forest cover.

Forests cover 48 percent of the total geographical area of the State (excluding Ladakh region) which is above the national average of 24.47 percent. It is, however, less than the norm of National Forest Policy, 1988, which provides forest cover of one third of total geographical area (20% for plains and 66% for hilly areas). The specie-wise area under forests has remained stagnant during the last few years. An area of 1075 Sq km is under Deodar constituting 5.31% of the total forest area. 1825 Sq km are under Chir having 9.02% share, 1969 Sq km are under Kail which accounts for 9.74%, 3401 Sq km are under Fir, representing 16.81%. Largest area of

10075 Sq km falls under other categories having a share of 49.8%. Thus, 18345 Sq km are under coniferous (Soft wood). Non-coniferous (hard wood) covers 1885 Sq km claiming 9.32% of total forest area. An area of 1962 Sq km is covered by Rakh and wildlife reserves.

Industry

The Industries sector is now playing its due role in the economic development of the State and in generating employment opportunities for the unemployed youth. Industries sector contributes 25.87% to GSDP of State at constant 2004-05 prices as per advance estimates for the year 2014-15. The State has 53 existing industrial estates spread over an area of 31335 Kanals of land as per position ending November, 2014.

The J&K State under the "Make in India " programme is looking forward to attracting domestic investors to invest in the State. "Ease of Doing Business" has rightly been recognized as the most important factor in achieving the goal of "Make in India". The State Government has initiated the process for playing its role as the interface of enterprises. The existing State Industrial Policy is expiring on 31st March, 2015 and the new Policy is going to come into effect from 1st April 2015. Committee constituted for drafting the new Industrial Policy has been properly sensitised to deliberate on the issue of easing out the procedures for the entrepreneurs/ industrialists to invest in the State. The Government is simultaneously working on drafting Trade Policy and manufacturing and Export strategies of the State.

Brief scenario of sericulture sector

Sericulture is the traditional occupation for a large section of the population (25,280 families) in the State. Inadequacy of mulberry leaves and damage caused by insects and pests has caused problems in the development and expansion of mulberry trees. Despite several initiatives taken by the State government, the number of mulberry trees is still very low (increased from 601, 000 to 1,402, 000 trees in the last 20 years). There is inconsistency in the production of cocoons due to the climatic limitation of the State and inadequate rearing equipment is another matter of concern. As a result, proper growth and development of silkworms does not take place resulting in decline in this profession.

Sericulture continues to be a subsidiary occupation for about 30,000 rural families in the State. Most of these families belong to economically backward sections of the society. Annually about 1022 MT of cocoons are produced generating an income of about Rs. 2026.00 Lac for these silkworm rearers coupled with the annual employment generation to the tune of 3.5 Lac man-days (3.0 Lac on-farm and 0.50 Lac 11 off-farm).

The Department has 173 mulberry nurseries spread over an area of 963 acrs, and 374 mulberry blocks over an area 2215 acres across the State. The nurseries are utilized for production of saplings/cuttings for further distribution amongst farmers to augment area under mulberry and mulberry blocks which serve as leaf reservoirs for the landless and marginal farmers. Annually about 20.00 Lac plants are produced from these nurseries against the potential of 30.00 Lac plants. The Department has a well-established seed organization and presently about 60% local annual seed demand is met out from the sector.

Cropping intensity

Cropping intensity is highest in Jammu region (176.8%) followed by Kashmir (123%) and Ladakh (106%) region. Though the cropping pattern is dominated by food crops, the State is deficient in food grains. The import of food grains year after year is suggestive of the fact that modernization of agriculture is imperative if the State has to achieve self-sufficiency in food grain production and ensure food and nutritional security of its people. It is to be recognised that food grain requirement for the estimated present population is of the order of 1.97 million t, whereas the production is 1.62 million t, a net deficit of 0.35 million t.

Status of mechanization

The small land holding, undulating topography, terraced irregular shape fields and low investment capacity of farmers makes the mechanization difficult with farm equipment available in the market. The traditional tools and implements are still in use with farmers. There are no specialized markets for the production and marketing of farm implements in the State except some manufacturing and marketing units in Jammu. Traditional hand tools and implements are made and maintained by local craftsmen. Divisional level Govt. Agricultural Engineering workshops exist in the State which manufacture and market subsidised hand tools and implements for farmers through various programmes of the State Government. The availability of agricultural workers, draught animals, diesel/electric pump, tractors, wheat thresher, ploughs and chaff cutter were 93.3, 141.6, 1.84, 1.25, 0.20, 732.9 and 8.79 per '00' ha gross cropped area, respectively (Sharma, 2009).

Constraints

Agriculture in Jammu and Kashmir has a significant history. The economic contribution of Agriculture to states GDP is steadily declining, still, agriculture is demographically the broadest economic sector and plays a significant role in the overall socio-economic fabric of the state. Boosting

Agricultural growth is very essential for inclusive growth because this sector sustains the livelihood of 70 percent of the population, However, agriculture contributes only 14.85 percent to states Gross Domestic Product. Low contribution of agriculture to GDP is 15% mainly because productivity of our major agricultural crops is very low in comparison to other states and the rest of the country. Technology deficit in agriculture seems to be the important cause. Yield per hectare of food grains, fruits, vegetables are far below the country level averages. Yield rates of Principal crops rice, maize and wheat for the year 2016-17 are 2550 kg/ha, 2509kg/ha and 3216 Kg/ha for rice, maize and wheat as per advance estimates for the year 2016-17 for India.

Looking at the present scenario, the area under the food grain crops in Jammu and Kashmir is continuously decreasing as the state is not able to feed its own people. According to the provisional figures of census 2011; the state of Jammu and Kashmir has a population of 12.54 million for which the per capita consumption of food grain is not available on its own. Rice and Wheat being the most consumed food grain crops is subject to different constraints as the area under rice is on a decline because people are shifting to other commercial crops which has compelled them to import from the rest of the states/nation. As per the data (right from 1947) the state Jammu and Kashmir has never been able to produce up to an extent so that it is going to feed its own people in terms of food grains as a result the gap between requirement and deficit is increasing.

The overriding challenge is sustainable and profitable growth in agriculture. Lack of political will has been the single most important constraint so far with a strong government now in office, there is reason to hope for gradual improvement. State is facing the low productivity of all agricultural crops and there is a massive deficit in its own production of food grains (forty percent), oil seeds (seventy percent) and vegetables (thirty percent). The net sown area of 7.52 lakh hectares (2004-2005) is thirty five percent of the reported area as against national average of forty six percent. Over seventy percent of the Net Sown Area is under food crops and the area under fruits is a little over thirteen percent. Viability of agriculture as a profession is presently affected by capital inadequacy, lack of infrastructural support and controls on movement, storage and sale etc. of agricultural produce. Dwindling water resources too is a major challenge as only forty two percent of the cultivated area is under irrigation. The lack of new technologies after the effects of the Green Revolution have worn off, in contemporary scenarios is becoming worrisome, only few promising seeds have been commercialised and most other innovations are still languishing due to poor extension or lack of investment.

Furthermore, corrective action regarding fertilizer subsidy, inadequate seed production, market rigidities and other market-distorting policies is desperately needed – but as yet political commitment is not visible. Achievement of irrigation potential is unlikely to surpass fifty percent of targeted acreage, given the large backlog of previous projects to be completed and the fact that no tangible change has been made in the strategy hence, monsoon dependence and erratic growth will continue to plague the sector. The effectiveness of government measures/schemes will remain erratic across the country, private participation can help but this will happen very slowly (fundamental enabling measures such as land reforms, corporations, taxation of agriculture do not look likely at this time) Fundamental degradation in environmental parameters and lack of Good Agricultural Practices (GAP) are not likely to be addressed in the foreseeable future, this will create a serious long-term threat. Soil degradation is approaching worrisome levels in most parts of India, fifty percent of total land and sixty-six of cultivated land is degraded – the highest amongst Asia Pacific countries. Water scarcity is projected as the single biggest factor for civil and social strife in the next decade.

Though the large diversity in agro-climatic conditions of the state is conducive for propagation of a diversified farming system, the terrain at the same time is tough and accessibility to a greater part of the region is poor. This causes a lot of hurdles to the inhabitants in provision of inputs, products and level of literacy and awareness. Climate change presents a real and exogenous force that will work against productivity improvement measures locally as well. India's overall crop yields could fall by thirty percent by 2050 according to the Intergovernmental Panel on Climate Change (IPCC) because of coastal flooding, greater drought incidence and reduced water availability. Other studies suggest a vulnerability of 5-15% in rich yields and 25-42% in wheat yields, response times of mitigation measures are 5-15 years.

In fact, the impact of climate change on the environment is visibly obvious and the valley is showing the incidents of the ill-effects of climate change. The performance of the agriculture sector depends on several drivers, which rather than impacting the sector isolation, interact with each other and also depend on sub-drivers, consequently strengthening or weakening specific trends. The key drivers that directly impact output can be grouped into six categories:

1. Technology (farming and crop technology)
2. Government policy (availability of credit, crop specific programmes etc.)
3. Cropping pattern (which depends on profitability, awareness etc.)
4. Environmental factors (water availability, soil degradation, climate change etc.)

5. Market forces (market openness, pricing, transparency, integration with downstream sectors)
6. Global factors (supply, demand, trade norms and restrictions)

(Maqbool and Jan, 2017)

Constraints of agricultural mechanization

The choice, extent and adoption of mechanization for Jammu and Kashmir agriculture have some severe limitations mentioned below:

1) The most crucial is the economic conditions of the farmers as the majority of the farmers are marginal and small.
2) The undulating topography and terraced irregular shape fields make mechanization difficult with farm equipment available in the market.
3) The availability and access to farm machines is very low due to lack of production and marketing facilities for farm implements in the State except Jammu region.
4) The packages of improved matching equipment with power tiller for agricultural operations have yet to be accepted by farmers of the region.
5) Custom hiring/ contract farming has not been emphasized for commercialization of agriculture. (Dixit *et al.*, 2014)

The Agri-Allied sectors face certain challenges and issues which are listed here.

Livestock sector

1. **Low productivity:** there is need to enhance the levels through genetic improvement along with practices such as; improving feed utilization efficiency, adopting better reproductive strategies and improving health coverage based on newer generation biotechnological vaccines and drugs.
2. **Shortage of feed and fodder:** Crop residue is used as animal feed but a large portion is burnt in the field to clear the field after harvest. Area under fodder cultivation is only about 4% of the cropping area and it has remained static over last four decades
3. **Unorganised:** Livestock sector particularly small ruminants; sheep, goat, piggery is highly unorganised.
4. **Inadequate infrastructure for marketing, processing and value addition:** Marketing of livestock and livestock products remains largely unorganised, traditional, and fragmented, with a few exceptions.

5. **Livestock and environment:** Climate change aggravates heat stress in dairy animals, adversely affecting their productive and reproductive performance. The estimated annual loss at present due to heat stress at the all-India level is 1.8 million tonnes. Livestock are itself a large source of methane emission contributing about 18% of total enteric methane budget.

Fisheries and Aquaculture sector

Challenges for fisheries development in India include the following:

1. Inland fishery not treated at par with agriculture in the context of taxes, electricity tariff etc. and therefore fishery sector remained largely unorganised and traditional in most parts of the country with little proliferation of technological improvements
2. Absence of inland fishery policy at national level
3. Non-coverage of fish farming under insurance
4. Lack of a reliable database relating to aquatic and fisheries resources
5. Non-availability of suitable fish yield models for multi-species fisheries for open inland waters and marine resources
6. Weak multi-disciplinary approach in fisheries and aquaculture
7. Inadequate attention to environmental, economical, social and gender issues in fisheries and aquaculture.
8. Inadequate human resource development and specialised manpower in different disciplines
9. Weak linkages between research and development machinery
10. Weak and unorganised marketing
11. Poor extension services to the fisher communities for promotion of fish seed and feed production, fish farming and fish-based enterprises
12. Poor technology transfer and anthropogenic interventions, resulting in loss of biodiversity
13. Decline in fish catch, depletion of natural resources due to over-exploitation of coastal fisheries
14. Pollution of water bodies with industrial and domestic effluents
15. Clandestine introduction and spread of exotic fish species
16. Unscientific management of fisheries and aquaculture activities
17. Contamination of indigenous fish germplasm resources
18. Poor yield optimization, problems in harvest and post-harvest operations, landing and berthing facilities for fishing vessels and issues in the welfare of fishermen.
19. Lack of value addition for enhancing profit margin

20. Lack of location specific improved technology
21. Lack of timely availability of inputs nearby

(Mishra, 2012)

Horticulture sector

Challenges for horticulture development in India include the following:

1. Lack of quality inputs
2. Lack of market support
3. Paucity of post-harvest management, packing and storage, specialised transport and storage arrangements to maintain the chain from farm to fork
4. Increasing cost of production; most growers do not get reasonable returns for their produce
5. Distress sales
6. Inadequate infrastructure like transportation, cold storage, warehouses etc.
7. Price fluctuation
8. Lack of market intelligence
9. Lack of knowledge of post-harvest handling
10. Wastage and spoilage loss
11. Lack of mechanization due to small land holding

Extension Services

Horticulture extension has been an integral part of agricultural extension services. During the last few years there has been diversification of agriculture towards high value commodities such as fruits and vegetables. Considering the evolving challenges, producers currently need a wider range of support - organizational, marketing, technological, financial and entrepreneurial.

Deterioration of quality of agricultural extension services in general and horticulture extension services in particular is observed, with experts and extension workers finding themselves lacking in capacity to recommend technology solutions to the producers. Therefore, there is a need to strengthen the technology support system.

Most of the extension personnel perform multiple roles. Their visits to the field are irregular as the service is preoccupied with the implementation of government schemes linked to subsidies and subsidized inputs.

In the State Directorates of Horticulture, administrative functions are generally assumed by the extension functionaries positioned at higher levels along the hierarchy line and extension work is left to the lower level / lowest functionaries. Most of these functionaries have academic qualifications of agriculture / horticulture but do not have direct exposure to farming; as a result, they lack confidence in addressing field level problems faced by farmers.

Sericulture sector

The sericulture industry is currently facing several problems which have restricted full utilization of its potential.

1. Silk yarn is of poor quality, which not only affects our competitiveness in the world market, but has also resulted in a preference for imported yarn in the domestic market. The problem arises due to lack of:
   - Sufficient thrust on the adoption of improved technologies;
   - Strict disease control measures;
   - Quality leaf due to insufficient inputs to mulberry garden;
   - Grading system for cocoons and
   - Quality-based pricing system as well as use of young age silkworms
2. Decline in area under silk food plants: This can be addressed by initiating area-specific research to improve soil fertility which will ultimately enhance soil productivity, increase mulberry and non-mulberry host plant leaf and silkworm cocoon production as well as arrest decline in area under silk food plants.
3. Insufficient production of bivoltine silk: Bivoltine yarn is sturdier and is used by the power loom industry.
4. The other factors responsible for it are;
   a) Insufficient adoption and proliferation of technology packages developed through research and development efforts
   b) Fragmented and ad hoc approach
   c) Non-involvement of private partners in a big way in seed production
   d) Non-penetration of the schemes
   e) Improper forward and backward linkages
   f) Dumping of cheap Chinese raw silk and fabric

Conclusions and suggestions

On the whole Jammu and Kashmir has huge potential to adopt agricultural mechanization for increasing agricultural production. Taking the fragile nature of the State and the conservation issues in sustainable farming into consideration, adoption of mechanization input to agriculture needs to be selected wisely. The following points are suggested as accelerate the pace of mechanization:

1) Hand tool technology can be improved for marginal and small farmers through introduction of appropriate and suitable tools developed elsewhere.
2) Adaptation of improved and suitable animal drawn matching implements and by using animals for other use to increase their annual utilization.
3) Popularization of mechanical power technology through large scale demonstrations on farmer's fields.
4) Development of custom hiring services in power equipment and machinery through entrepreneurs, self-help groups and cooperatives.
5) Creation of repair facility with a supply of genuine spare parts of farm machinery.
6) Development of a proper agricultural engineering package for the State.
7) There is scope for micro-irrigation systems largely due to the scarcity of irrigable water and uneven terrain conditions. (Dixit *et al.*, 2014)

A number of private seed companies, pesticide and insecticide companies provide extension services and transfer production and post-harvest management technologies to producer farmers. However, due to lack of technical knowledge of products this may result in indiscriminate use of farm chemicals.

Buyers who enter into contract production programs of horticulture crops for exports, processing or domestic marketing do supply seed and planting materials, other farm inputs and relevant technologies. They may also provide cold chain and packaging solutions for long distance transport of produce.

References

Dixit, J., Jeena, A. S., Shahi, N. C. and Wahid. T. 2006. Potential of farm mechanization in Jammu and Kashmir State of India-A review. *Agricultural Mechanization in Asia, Africa and Latin America* **37**(4): 9-14.

Dixit, J., Sharma, S. and Ali, M. 2014. Present status, potential and future needs for mechanization of agricultural operations in Jammu and Kashmir state of India. *International Agricultural Engineering Journal* **16**(3): 87.

Kaloo, M. J. and Choure, P. T. 2021. Present Status and Future Prospects of Agriculture in Jammu and Kashmir. *IOSR Journal of Humanities and Social Science (IOSR-JHSS)* **20** (11): 62-67.

Maqbool. O and Jan. F. 2017. The Agricultural Sector of Kashmir: Between Growth and Mismanagement. *Academic Research Journal of Agricultural Science and Research* **5**(5): 341-349.

Mishra, N. 2012. A study on the issues of the inland fisheries and the policy framework governing the sector. Available at; https://www.indiawaterportal.org/articles/study-.issues-inland-fisheries-and-policy-framework-governing-sector.

Sharma, S. 2009. Status of farm power availability in the selected villages of Jammu District (J&K State). *Agricultural Mechanization in Asia, Africa and Latin America* **40**(10): 41-45.

Sharma, S. 2010. Status of farm power availability in the selected villages of Udhampur District (J&K State). *Journal of Dryland Agriculture Research & Development* **25**(1): 70-73.

Chapter 2

Impact of Urban Green Spaces on Mental Health

Salma Farooq, Shah Murtaza Mushtaq, Amir Bhat and Akhlaq Amin Wani*

Division of Natural Resource Management, Faculty of Forestry SKUAST- Kashmir

**salmafarooqui027@gmail.com*

Introduction

Green space in cities shouldn't be considered an optional luxury. It is a crucial part of a healthy human habitat- Charles Montgomery

Green Spaces: The WHO defined Urban Green Spaces as all urban land covered by vegetation of any kind and includes parks, playgrounds and residential greenery. In 2017, WHO presented eight classes of green spaces: roadside greenery and other vegetation barriers that could be found along streets or rail tracks; small urban green spaces (such as gardens or pocket parks) and playgrounds; green roofs and facades; parks and urban meadows; greenways and corridors (such as green trails for walking/cycling); green with blue spaces (that could be found in coastal, riverside or lakeside trails); recreational and urban gardening facilities (such as community gardens, sport and play areas and school grounds); and finally the facilitated access to urban woodlands, forests and natural wildlife areas. A set of dimensions could characterise green spaces; their location, distance to users, size, quality and security, all aspects that could be aggregated in an "availability and accessibility" dimension. Besides, there are also aspects related with landscape quality and the user's perception, designated as "aesthetic" dimension, that include aspects like the type of services provided by green spaces and the possibilities of management of these spaces. (Costa and Kallay, 2020)

LEAVING NATURE: Over the last 100 years, our increasingly urban lifestyles (75% of the European population now live in cities and towns) have disconnected us from nature and this may have contributed to a decline in many aspects of our health and wellbeing. The UK Department of Health reported that children now spend only 9% of their time outdoors, while for adults the figure is approximately 20%. Indoor lifestyles are often associated with reduced exercise, increasing rates of obesity and diabetes, leading to higher incidences of depression and other psychiatric disorders. Estimates of the cost to the UK National Health Service (NHS) of reduced physical activity are £1-2 billion per annum (this equates to$1.6-3.2 billion per annum; date of conversion March 26, 2011). The Biophilia Hypothesis has yet to be subjected to rigorous scientific analysis, and there may well be other reasons for the increasing incidences of some of the disorders mentioned above. However, a weight of evidence is accumulating that natural environments have an important influence on our lives. Time spent in nature affects levels of physical activity resulting in reduced health service costs and also in the number of individuals claiming to have experienced the personal benefits of spending time in forests, on seashores, on river banks, mountains, or moorland. Barton and Pretty explored the dose response relationship between humans and natural environments. Exercise in the presence of nature (so-called green exercise) leads to positive short- and long-term health outcomes.

Nature in our Living Spaces

Globally, more and more people are living in cities. Also, over the last 20 years, fewer people have been visiting the country side. This makes it especially important that towns and cities include within their precincts opportunities to come into contact with are as rich in vegetation (green spaces) and aquatic environments (rivers, canals, ponds, lakes, the sea coast—so-called blue spaces). A U.S. study in Philadelphia suggested that maintaining city parks could achieve yearly savings of approximately $69.4 million in health care costs. Given the renewed interest of several national governments in reducing the health costs associated with illness in their populations, it seems timely that we initiate new research to determine just what characteristics of natural environments are beneficial to humans. Improved understanding could also be helpful to those who, for one reason or another, do not have easy access to nature; in particular, the elderly and infirm who spend almost all of their time indoors.

Understanding how Natural Environments can be used to Promote better Health

There already exists a quite extensive literature regarding opportunities to use nature to promote health and wellbeing. Much of the work is anecdotal however, or involves small-scale studies, often lacking appropriate controls or statistical robustness. There are exceptions though. As was alluded to earlier, Attention Restoration Theory has been built upon evidence that time spent in natural outdoor environments powerfully assists recovery from mental fatigue. Work of this kind has led others to instigate programs that facilitate contact with, and physical activity within, natural environments as a way of tackling major disease issues such as obesity-related diabetes and depression, as well as improving general health and wellbeing. Bird reviewed the literature and identified evidence that time spent in natural environments was beneficial for improving mental health in people of all ages, improving self-discipline in children, helping the treatment of attention deficit hyperactivity disorder, reducing aggression and crime, promoting better health in the elderly, and delaying the impact of Alzheimer's disease. Practical measures that take advantage of this knowledge include, for example, the Green Gym program that was initiated in 1997 with a view to encouraging people to become more active by engaging in gardening or conservation activities. The Green Gym Website offers a "virtual green gym" tour to introduce the program. More recently, this approach was adapted to launch the Blue Gym program. Here, activities in the coastal marine environment ranging from mild activity associated with rock pool rambles and coastal walks, to more vigorous pursuits such as swimming, sailing, kayaking, and surfing, are being used to motivate people to spend more time outdoors engaging in physical activity. The Blue Gym has built into its operating structure rigorous scientific research led by the Peninsula College of Medicine and Dentistry (Universities of Exeter and Plymouth, UK) to ensure that changes in health and wellbeing can be assessed objectively. Elsewhere, research by Tsunet sugu and colleagues from the Japanese Forestry and Forest Products Research Institute has addressed the exposure of subjects to forest and urban settings and the effect that these exposures have on subjective ratings and physiological measures, including blood pressure, heart rate, and salivary cortisol excretion. Their studies demonstrated that 15 to 20 min exposures to natural environments, such as a broad leaf forest, were accompanied by significant lowering of blood pressure, pulse rate, and cortisol levels when compared to similar exposures in a busy city area. Subjective ratings of "calm","comfortable", and "refreshed" were higher in forest conditions than those recorded in the city. Others have also recorded reductions in blood pressure and cortisol following time spent in natural environments. Such studies suggest that a reduction in symptoms related to prolonged stress and depression can be achieved, including those brought about by long periods of bad weather and annual time changes—forms of" constrained restoration". There are numerous other examples.

Watching animal movement, admiring flowers and trees, listening to birds and other outdoor sounds, viewing spectacular cloud formations or sunsets, and the familiar natural odors are among the diverse attributes of nature that in some way promote wellbeing.

Natural environment i.e., green spaces and well-being are intrinsically linked with each other has positive effects on physical, mental and social well-being, Environmental pollution is one of the serious problems faced by the people in the country. Rapid population growth, industrialization and urbanization in country are adversely affecting the environment. Though the relationship is complex, population size and growth tend to expand and accelerate these human impacts on the environment. All these in turn lead to an increase in the pollution levels. However, environmental pollution not only leads to deteriorating environmental conditions but also have adverse effects on the health of people. India is one of the most degraded environment countries in the world and it is paying heavy health and economic price for it. The present paper is an attempt to examine population growth, increasing urbanization and its influence on the environment and health of the people. The secondary analysis conducted of changes and trends over last fifty years. The analysis reveals that rapid population growth plays an important role in environmental problems of the country, from deforestation to land degradation, air and water pollution to the spread of disease. The considerable magnitude of air and water pollution pulls up the number of people suffering from respiratory and water borne diseases and many a times leading to deaths and serious health hazards. The analysis suggests that there is a urgent need to control population and environmental pollution in the country for better health of present and future generation.

Some of the effects of natural environment on physical health include:

- Healing and restorative effects
- Promotes physical activity and health
- Combates obesity

Effects on mental well-being:

- Recovery from mental illness
- Relieve stress, fatigue, anxiety and depression
- Improving mood

Effects on social well-being:

- Promotes cohesion
- Promotes social interaction
- Encourages sense of place
- Feelings of safety and security

(Newton, 2007)

Health effects of urban built environment

Urban environment directly affects the health and quality of life of the urban population. According to the estimates of the World Health Organization (WHO) 63% of global mortality, about 36 million deaths per year, is due to chronic diseases. A large proportion of these deaths are associated with risks related to the urban built environment, such as physical inactivity and obesity, cardiovascular and respiratory diseases due to transport-generated urban air pollution, and heat-related strokes and illnesses. Globally, household and outdoor air pollution is responsible for 7 million premature deaths each year, out of which outdoor air pollution is causing 2.8 million deaths attributed to chronic diseases. The urban heat island effect aggravates heat stress during heat waves that can lead to fatal heat stroke. As the effects of climatechange accelerate, and related heat waves become more frequent, the heat island effect will become even more pronounced. Physical inactivity, that is likely to be more common among urban populations (due to poor walkability and lack of access to recreational areas) is responsible globally for 3.2 million deaths annually (Bennett and jones, 2018).

Health impacts of urban green spaces

Academic literature, the World Health Organization (WHO) and other institutional sources highlighted the importance of green spaces to improve health and well-being. WHO presented a causal model of the impacts of urban green spaces on health and well-being. Through improved air and water quality, reducing noise levels and contributing to temperature regulation, urban green spaces can effectively reduce environmental health risks associated with urban living. In addition, they deliver health and well-being by enabling stress reduction and relaxation, physical activity, improved social interaction and community cohesion. Access to natural environments can improve overall mental health, physical fitness level, cognitive and immune function, and can lower mortality rates in general. (Costa and Kallay, 2020)

Impacts on physical health

There is evidence that the provision of open and green spaces is associated with improved general physical health outcomes. According to the European Environment Agency every 10% increase in green space is associated with a reduction in diseases equivalent to an increase of five years of life expectancy.

A systematic review of five online databases and over 100 studies undertaken by the University of East Anglia found that people who spend more time in green spaces have significantly reduced risks for a number of chronic illnesses. According to the research, exposure to green spaces was linked to lower heart rate, lower blood pressure, lower 7 cholesterol, and reduced incidence of stroke, asthma, diabetes and coronary heart disease.

Some studies have indicated that there is a positive correlation between higher levels of green space and lower levels of obesity.

Evidence from a systematic review of studies shows that living in areas with higher residential greenness reduces the risk of cardiovascular mortality. In Florida, higher level of exposure to green space has been associated with a reduced risk of stroke mortality.

A study undertaken in Australia was highlighting the relevant role of the diversity of landscape features, such as trees and open spaces in improving physical health outcomes. The study has indicated that a greater variability of neighbourhood greenness was associated with reduced risk of cardiovascular disease and stroke. The likelihood of hospitalization and of self-reported heart disease or stroke was lower among people who lived in urban areas with highly variable greenness. It is suggested that heterogeneity in the distribution of greenness might contribute to better health outcomes through promoting physical activity.

Direct exposure to green space improves the functioning of the immune system. According to a Finnish study, the immune systems of children can strengthen in a short time if their day-care centre yards are made more natural using forest undergrowth and grass. The findings suggest that exposure to natural environment can change the microbiome of children that result in better functioning of their immune systems.

Residential access to green space was also found to be linked to improved pregnancy outcomes. In the 1980s, the study of Roger Ulrich showed that merely having a view of nature can provide physical benefits. Patients with a view of trees from their hospital bed recovered more quickly, required less pain medication, and had fewer post-surgical complications than patients in rooms with urban views (Berg *et al.,* 2015).

Impacts on mental health

Access to natural environments can also improve overall mental health. There is an indication that experiencing the natural environment reduces stress levels.18 The impact of green spaces to mental health also include improved general mood, reduced depressive symptoms, enhanced cognitive functioning, improved mindfulness, short-term memory performance and enhanced creativity.

At individual level, the lack of contact with natural spaces is linked to an increase in the incidence of mental illnesses .Based on a growing body of evidence suggesting that living without nature is unhealthy, Richard Louv has coined the term '**nature deficit disorder**'

Nature Deficit Disorder

Over the last century, urbanization has become one of the most important topics that have effects on several aspects of citizens' health. According to the World Bank report, 68% of the world's population will live in cities by 2050. This change in lifestyle of humanity has several effects on urban health, which include urban governance; population characteristics; the natural and built environment; social and economic development; service and health emergency management; and food security. The characteristics of urban open spaces have specific impacts on citizens' health. The population growth in developing countries, especially in Asian countries, has caused a decreasing rate of open spaces per person in recent years and it could be worse in the future. The impact of natural areas on children's health and how it would be effective on Nature-Deficit Disorder treatments, which cause several mental and physical issues for children, is the focus of this study. The main idea of this study is based on the Nature-Deficit Disorder concept, which has discussed by Richard Louv regarding the importance of nature connection on children's wellbeing. In this study, an in-depth interview was conducted with primary school students

– children between age 4 and 12 – and their teachers in Mazandaran, the northern and green province of Iran, and the educational environment has been selected as a tool to measure the impacts of being in nature as a daily activity of children's wellbeing and their learning quality. As a result of the research, we found that students with more daily nature connection at school or in their neighbourhoods have better performance and capability at school and also in their social and cognitive skills.

Urban health and well-being

A rapid rise of urbanization and the intricacy of urban structures and the complexity of diverse factors have made the urban health one of the hottest topics all around the world. At first, it is necessary to take a quick look at the definition of urban health and well-being. Urban health includes a group of social and physical features that are the results of urban infrastructures that have impacts on public health and quality of life in urban areas. Well-being is a concept with three main elements; happiness, life satisfaction, and quality of life. The combination of urban health and well-being shapes the concept of urban wealth, which means the overall level of quality of life and satisfaction of a city. The quality of urban wealth is determined by several factors such as individual physical, mental, emotional aspects of health. An unhealthy urban environment has negative impacts on citizens, such as; physical health problems, mental illnesses, reduction of productivity, increasing the general stress level. Urban green/open spaces, on the other hand, have therapeutic impacts for citizens; they improve mental and physical health and also reduce urban environmental quality problems like air and noise pollutions in cities. Since the 19th century, during industrialization and the rapid growth of urbanization, urban health becomes one of the growing concerns in urban studies. Urban health is measured by several indicators such as; physical, biological, chemical, and social factors, but in spatial planning, the main factor to measure the level of quality of life in the urban environment is related to the accessibility to green areas.

- Not a clinical diagnosis
- Nobody will be hospitalized with nature deficit disorder
- A metaphor of the fact that there are growing scientific evidence that living without nature is unhealthy
- A call to rethink and discriminalize the time spent in nature
- Richard louv (2005): Last child in the wood.

Effects of Nature Deficit Disorder: short term

- Strong scientific evidence from a metaanalysis of studies:
- Short-term exposure to forests, urban parks, gardens and other semi-natural environments has been found to:
- Reduce stress and depressive symptoms,
- Increase self-reported positive emotions,
- Improve self-esteem, mood and preceived mental and physical health
- Produce physiological and psychological relaxation
- Reduce hyperactivity and attention disorders in children

Effects of Nature Deficit Disorder: long term

- Strong scientific evidence from a meta analysis of studies:
- Long term exposure to natural environments, such as residing in areas with high greenness or in diverse landscape, has been associated to reduced mortality and to improved mental health

(Richard louv, 2005)

Benefits of green spaces in cities

- WHO said that physical inactivity as a major public health risk
- Studies found -people using open spaces-three times more physically active
- Improving access-creating network of parks- green spaces results in:-
- Increased physical activity
- Improved mental health
- Better general health
- Reduced stress levels
- Reduced depression
- Reduced healthcare costs
- Environment benefits
- Providing healthy habitats for humans, wildlife, plants in densely built places
- Preserves ecosystems aimed growing cities

(Lee and Maheswaran, 2011)

Theories associated with mental health and nature

There are several theories that attempt to explain the beneficial effects of nature exposure:

- **Ulrich`s Stress Reduction Theory (SRT)** proposes that contact with unthreatening natural environments or viewing natural elements, having a restorative effect can be effective in reducing stress creating positive emotions and feelings (Ulrich, 1981)
- **Kaplan's Attention Restoration Theory (ART)** suggests that continued attention and fatigue can degrade our ability to solve problems and cause various negative emotions, and that mental fatigue and concentration can be improved by time spent in, or looking at nature, showing that natural environment can resolve health problems of modern people's daily lives caused by stress and fatigue. Nature exposure allows individuals to be away from daily stressors, as natural environment attracts their attention through 'soft fascination' providing the opportunity for recovery from mental fatigue.
- **Wilson's biophilia hypothesis** claims that, as a consequence of evolution, humans have an innate tendency to interact positively with nature. (Kellert and Wilson, 1993)

Mental Health

Mental health refers to cognitive, behavioral, and emotional well-being. It is all about how people think, feel, and behave. People sometimes use the term "mental health" to mean the absence of a mental disorder. "Mental health is a state of well-being in which an individual realizes his or her own abilities, can cope with the normal stresses of life, can work productively, and is able to make a contribution to his or her community. Mental illnes or mental health disorders are a wide range of conditions affecting mental health. mental disorders affect mood, thinking and behaviour of individuals. Some common mental disorders includes anxiety, depression, eating disorders (anorexia and bulimia) and addiction. In most cases, the symptoms of mental disorders can be managed conservatively with medications and psychotherapy. Mental illness can occur at any age, as a child or as an adult (Prince *et al.*, 2007).

Signs of mental illness

Long lasting sadness or irritability

Extreme changes in moods

Social withdrawal

Dramatic changes in eating or sleeping pattern
Inability to concentrate
Excessive fears
Unnecessary feelings of guilt
Easy irritability
The tendency to react with voilence
Withdrawal from friends and activities
Inability to work or study

(Prince *et al.*, 2007)

Green Prescriptions

The 'Green Prescription' term introduced in **New Zealand** in **1998**. A green prescription is an emerging nature-based health intervention, that is designed to improve physical and mental health and wellbeing of patients through exposure to natural environments. It is a written advice of a health professional to a patient to be physically active. the doctor recommend the patient to go for regular jogs or walks in the park, or to participate in some other nature-based activity. It is a smart and cost-effective option to promote health creation. It includes activities: regular group walks; gardening ; biodiversity conservation activities, such as habitat creation and restoration; green exercise (e.g. nature walks, biking, climbing) Robinson and Breed, 2019.

Initiatives

The trend for urban greening has already begun, pointing some inspiring examples in the US, including **the Million Trees Los Angeles** initiative and an ambitious greening programme in New York. This isn't just a richer-world phenomenon, either. Most urban growth in the next decades will occur in lower-income nations. **The Milan Urban Food policy** Pact, which aims to increase urban gardening around the world, has 211 cities signed up, many in Africa, South America and South-East Asia. **China's Ministry of Ecology and Environment** established in 2018, has made fighting pollution one of its three critical battles, spurring the building of parks, green spaces and wildlife corridors in many cities. Lower-income countries face many challenges in building greener cities but they can learn from the mistakes already made in lower-growth cities in west. There's a lot of opportunity for sustainability in developing cities (Douglas, 2021).

Recommendation

Some of the recommended interventions included are listed below by age group:-

- Children: engagement with nature and physical activity should be encouraged with certain designed-in elements, such as paved walkways, play equipment, fields and courts;
- Adolescents: provision of sports facilities and other facilities supporting movement and physical activity; provision urban wilderness areas, untended vegetated areas, wildflower meadows for exploring and creative play; and zones of sheltered areas for relaxation and social interaction;-
- Adulthood: ensuring proximity of green spaces; provision of opportunities to engage with nature, spaces for walking, cycling for stress relief; development of communal seating areas in parks to encourage social interactions;-
- Pregnant women: provision of safe and accessible walkways, regular seating areas and clean public toilets;
- Elderly: provision of accessible green spaces in close proximity to residential areas, sheltered seating areas with interesting views that encourage social interaction, spaces for games suitable for elderly abilities, such as chess and boules_x0002_type games; increasing exposure to green space through targeted greening of the streetscap.

Conclusion

"How we plan cities now affect the well-being of billions in the future"?

- Community involvement ensures that different cultural and local needs are met
- People need to also experience the effect that urban green spaces have on their sense of well-being
- Our planning today and into the future-will affect the well-being of billions of people
- Recent studies from China and England find that feeling more connected with nature makes people more likely to adopt positive environmental behaviours
- Advantages of developing urban green spaces is that their health promoting effects are democratic that is the effects are equal for men, women, across all age groups and across all socio-economic categories.
- Greener cities won't just improve the mental health of their residents, but also focus our minds on the needs of nature beyond our urban jungles.

References

Andrey *et al.,* 2016. Urban green spaces and health: WHO Regional Office for Europe.

Douglas, K. 2021. The nature fix. New Scientist, 249(3327), 36-40.

Kellert, S. R., and Wilson E. O., 1993. eds. The biophilia hypothesis. *Island Press.*

Lee, A. C., and Maheswaren, R. 2011. The health benefits of urban green spaces: a review of the evidence. Journal of public health, 33(2), 212-222.

Louv, Richard 2005. Last Child in the Woods. Saving Our Children from Nature-Deficit Disorder, New York: Workman Publishing Company

Maas, J., Van Dillen, S. M., Verheij, R. A., and Groenewegen, P. P. 2009a. Social contacts as a possible mechanism behind the relation between green space and health. Health & place, **15**(2): 586-595.

Maas, J., Verheij, R. A., de Vries, S., Spreeuwenberg, P., Schellevis, F. G., and Groenewegen, P. P. 2009b. Morbidity is related to a green living environment. *Journal of Epidemiology &Community Health*, **63**(12): 967-973.

Marques da Costa, E., and Kallay, T. 2020. Impacts of Green Spaces on Physical and Mental Health.

Netwon, J. 2007. Well-being and the natural environment: a brief overview of the evidence. University of Bath, UK.

Prince, M., Patel, V., Saxena, S., Maj, M., Maselko, J., Phillips, M. R., and Rehman, A. 2007. No health without mental health. *The Lancet*, **370**(9590): 859-877

Robinson, J. M. and Breed, M. 2019. Green Prescriptions and Their Co-Benefits: Integrative Strategies for Public and Environmental Health. Challenges. 10. 9. 10.3390/challe10010009.

Twohig-Bennett, C., Jones, A. 2018. The health benefits of the great outdoors: A systematic review and metaanalysis of greenspace exposure and health outcomes. Environmental Research. Volume 166, October 2018, Pages 628-637

Ulrich, R. S. 1981. Natural Versus Urban Scenes. Some Psychophysiological Effects. September 1981, Environment and Behavior, **13**(5):523-556.

Van den Berg, A. E. and Van den Berg, M. M. H. E. 2015. Health benefits of plants and green space: Establishing the evidence base. Acta Horticulturae, 1093, 19–30. https://doi.org/10.17660/ActaHortic.2015.1093.1

WHO, 2012. Action plan for implementation of the European strategy for the prevention and control of noncommunicable diseases 2012–2016. Copenhagen: WHO Regional Office for Europe.

WHO, 2013. Physical activity promotion in socially disadvantaged groups: principles for action. Copenhagen: WHO Regional Office for Europe.

WHO, Regional Office for Europe 2016. Urban green spaces and health. A review of evidence.

Chapter 3

Phytosociology of Gulmarg Forest Range of Kashmir Himalayas

Suraya Tasveer[1], Aasif Ali Gatoo[1], Akhlaq Amin Wani[1], M A Islam[1], Shah Murtaza Mushtaq[1] and P A Khan[2]*

Division of Natural Resource Management, Faculty of Forestry SKUAST- Kashmir[1]
Division of Forest Biology and Tree Improvement, Faculty of Forestry SKUAST- Kashmir[2]

**Email: asifskuast@gmail.com*

Introduction

The Kashmir Himalaya is well-known for its aesthetic beauty all over the world; and is referred to as a "terrestrial Paradise" on the earth (Vigne, 1842). One of the chief features contributing to the global fame of Kashmir is the rich biodiversity that adoms its captivating landscape (Lawrence, 1895).Being Phyto-geographically located at the intersection of Holarctic and Paleotropical Floristic Realms and falling within the North-Western Himalaya, the Kashmiri Himalaya region is endowed with teeming treasure troves of plant diversity. Due to varied topography and habitat heterogeneity along a wide elevation range it represents a unique bio-region. As supported by the geological evidences (Puri, 1943, 1947; Vishnu-Mittre, 1963), The Kashmir Himalayan flora, to begin with, is believed to have been typically a tropical one; and in the aftermath of Little Ice Age, it was transformed into a mosaic of subtropical-temperate-alpine flora. With the passage of time, the floristic elements from far and wide have found entry into this region, either through natural dispersal mechanisms or by human interferences; hence its extent flora is largely an admixture of neighbouring Photo-geographical regions with reasonably good proportion of endemics. Himalayas are well known for their diverse and characteristic vegetation distributed over a wide range of topographical variations. Western Himalayas are rich in tree diversity but the effects of human influences on the vegetation like tree felling and lopping, over grazing, surface mining, defense and developmental activities and road building activities are severe in the region and these activities have damaged the flora of other region very severely (Gupta, 2004). The temperate and sub-alpine forests of western Himalayas are also suffering from the vagaries of human influences as well as climatic changes (Gairola *et al.*, 2008). The Himalayas are the home of conifers and constitutes one of the maximum divergent and economically essential institutions of species.

Phytosociology is defined as the discipline which concerns itself with the study of vegetation as such, with its floristic composition, structure, development and distribution (Tansley, 1920). The study of species diversity is helpful in understanding community composition, structure, change and development (Li *et al.*, 2002). Species diversity is one of the most important indices used for evaluating the stability and sustainability of forest communities. Information on the species composition of a forest is essential for its wise management in terms of economic value, regeneration potential (Wyatt-Smith, 1987) and ultimately may be leading to conservation of biological diversity (Verma *et al.*, 1999). Phytosociological studies are crucial for protecting the natural plant communities and biodiversity as well as understanding the variations experienced in the past and continuing on into the future. Most developed countries have accomplished these basic studies (Hamzaoğlu,

2005) and primed vegetation maps (Tel *et al.*, 2010). A key component of biodiversity studies is the Phytosociological investigation done with the aid of sampling techniques and expressed quantitatively both in absolute terms of species with respect to all other plant species of the area. Species diversity in the tropics differs dramatically from place to place. Quantitative inventories help in identification of frugally useful species as well as species of special concern, i.e. rare, infrequent and vulnerable species and hence to quantify conservation worthiness of the candidate sites (Keel *et al.*, 1993). Phytosociology attempts to describe the diversity in plant communities and its methods often involve the quantitative assessment of various parameters of vegetation like cover, abundance, frequency etc. Phytosociological surveys are imperative tools of ecologists to assess and appraise the vegetation types of given ecosystem. The Phytosociological information about each distinct tree species is vital for understanding their ecology and establishing conservation management policies for over exploited forests (Kharakwal *et al.*, 2009). Phytosociology of Himalayan forests has been subject to extensive research (Gairola *et al.*, 2008). Although much work has been conducted in Eastern and Western moist temperate Himalayan forests, the Kashmir Siwaliks have not been given proper attention due to remoteness, lack of infrastructure and danger, being at India- Pakistan border (Shaheen *et al.*, 2011).

Species diversity is one of the most important indices used for evaluating the stability and sustainability of forest communities. Information on the species composition of a forest is essential for its wise management in terms of economic value, regeneration potential (Wyatt and Smith, 1987) and ultimately may be leading to conservation of biological diversity (Verma *et al.*, 1999). The study of species diversity is helpful to understanding community composition, structure, change and development (Li *et al.*, 2002). Moreover, species diversity is affected by multi-environmental factors (Gaston, 2000; Lan, 2003; Tang *et al.*, 2004) especially by the altitudinal gradient. The altitudinal gradient is an important factor affecting species composition and structure (Whittaker, 1972). Species diversity has functional consequences, because the number and kinds of species present in any area determine the organismal traits which influence ecosystem processes. Species diversity reflects species richness and evenness and expresses differences of community structure, community composition and community habitat conditions (Lan, 2003). Distinguishing plant communities has been at the heart of vegetation science for centuries, with a traditional focus on the distribution, composition and classification of plant communities (Kashian *et al.*, 2003)

Materials and methods

The present investigation was conducted in Four blocks of Gulmarg Forest Range *viz*; Khaipora, Ferozpora, Tangmarg and Baderkut in Union territory of Jammu and Kashmir during 2019-2020.The Gulmarg Forest range is situated between 34°3'31.2"N to 74°23'1.0"E. The total area of Gulmarg forest range is 15933.78 ha. The procedure of Stratified random Sampling was followed. The details of experimental plots, materials used and methodologies adopted for this study are discussed as:

Study site

Table 1: Details of the study area:

Block	Block Name	Elevation (m amsl)	Longitude	Latitude	Aspect	Slope
1	Khaipora	2345	34°03'15.9"N	74°24'40.0"E	NW	35°
2	Tangmarg	2265	34°03'40.93"N	74°25'15.58"E	NE	35°
3	Ferozpora	2365	34°2'55.59"N	74°28'50.98"E	SW	45°
4	Baderkut	2221	34°02'25.64"N	74°27'04.29"E	NE	45°

Phytosociology of Gulmarg Range

Vegetation analysis was carried out by stratified random sampling. Quadrats of 50× 20m, 5 × 5m and 1 × 1m for trees, shrubs and herbs, respectively were laid at each site. The vegetation data were computed for density, frequency and basal area following Curtis and McIntosh (1950). The relative values of frequency, density and dominance were determined as per Phillips (1959). These values were summed to represent Importance Value Index (IVI) of individual species in order to express the dominance and ecological success of the species (Curtis, 1959). The important quantitative analysis such as density, frequency, and basal area of tree species, shrubs and herbs species were determined as per Curtis and McIntosh (1950). Importance Value Index (IVI) was used to determine the overall importance of each species in the community structure. In calculating this index, the percentage values of the relative frequency, relative density and relative dominances were summed up together and this value were designated as the Importance Value Index or IVI of the species (Curtis, 1959).

Shannon diversity index (Ḣ)

Species diversity is determined by Shannon's index of general diversity and was calculated by using Margalef formula (Odum, 1971)

Shannon index of general diversity (Ḣ) = $-\sum(ni/N)\ \text{Log}\ (ni/N)$

Where,

ni – Importance value of each species; N – Total importance value

Experimental findings

Phytosociology

During the present investigation, a total of 54 plant species (Table 2) belonging to 25 families were recorded at different blocks of Gulmarg Forest range which include a total of 08 tree species, 11 shrub species, 34 herb species and 01 fern species. Out of the total families recorded, Asteraceae was the dominant family represented by 08 species followed by Poaceae with 6 species. Rosaceae and Pinaceae with 04 species each. Geraniaceae, Adoxaceae, Ranunculaceae and Polygenaceae with 02 species in each. While as Sapindaceae, Juglandaceae, Taxaceae, Hamamelidaceae, Apiaceae, Saxifragaceae, Onagraceae, Oxalidaceae, Phytolaccaceae, Lamiaceae, Utricaceae, Scrophulariceae, Violaceae and Pteridaceae were represented by only one species in each family.

Table 2: Species composition at different sites in Gulmarg Forest Range of Kashmir Himalayas

S.no	Species	Family	Common name	Local name	Tangmarg	Khaipora	Ferozpora	Baderkut
	Trees							
1	*Abies pindrow* spach	Pinaceae	Fir	Budul	+	+	+	+
2	*Aesculus Indica* (Wall.ex Cambss.)	Sapindaceae	Indian horse-chestnut	Handoon	+	-	-	-
3	*Cedrus deodara* (roxb.ex D.Don) g.Don	Pinaceae	Himalayan cedar	Deodar	+	-	-	+
4	*Juglans Regia* L.	Juglandaceae	English Wallnut	Doon	+	-	-	-
5	*Picea smithiana* (Wall.) Boiss.	Pinaceae	Himalayan spruce	Rai budul	+	+	+	+
6	*Pinus wallichiana* A.B.Jacks.	Pinaceae	Blue pine	Kayur	+	+	+	+
7	*Robinia pseudoacacia* L.	Fabaceae	Black locust	Kikar	+	-	-	-
8	*Taxus contorta* Griff	Taxaceae	Common yew	Poshtul	-	+	-	-
	Shrubs							
1	*Berberis aristata* DC.	Berberidaceae	Indian Barberry	Kawdach	+	+	-	-
2	*Berberis jaeschkeana* C.K.Schneid	Berberidaceae	Jaeschke's Barberry	Kawdach	-	+	+	-
3	*Berberis lycium* Royle	Berberidaceae	Indian Barberry	Kawdach	+	-	-	-
4	*Indigofera heterantha* Brandis	Fabaceae	Himalayan indigo indigo	Jand	-	-	+	+
5	*Parrotiopsis jacquemontiana* (Decne.) Rehder	Hamamelidaceae	Witch Hazel	Pohu/Hatab	-	-	+	+
6	*Rosa canina* L.	Rosaceae	Dog rose		+	-	-	-
7	*Rosa webbiana* Wall.ex Royle	Roseaceae	Wild rose	Shingai	-	-	-	+
8	*Rubus fructicosa* L.	Rosaceae	Raspberry	Chaanch	-	+	-	-
9	*Rubus niveus* Thunb.	Rosaceae	Hill raspberry	Chaanch	+	-	+	+
10	*Sambucus wightiana* Wall.ex Wight	Caprifoliaceae/ Adoxaceae	Elderberry	Fhakee	-	-	+	-
11	*Vibernum grandiflorum* Wall.ex DC.	Adoxaceae	Grand viburnum	Kulmaach	+	+	+	+

Table 2

S.no	Species	Family	Common name	Local name	Tangmarg	Khaipora	Ferozpora	Baderkut
	Herbs							
1	*Achillea millifloium* L.	Asteraceae	Old man's pepper	Pahal gasseh	+	+	+	-
2	*Aegopodium podagraria* L.	Apiaceae	Ground Elder		-	+	-	-
3	*Anthemis cotula* L.	Asteraceae	Stinking chamomile		-	+	-	-
4	*Bergenia ciliate* (Haw.) Sternb.	Saxifragaceae	Winter begonia	Zakhmi hayat	-	-	+	-
5	*Bothriochloa ischaemum* (L.) Keng	Poaceae	Yellow stem		+	-	-	+
6	*Centaurea iberica* Trevir.ex Spreng.	Asteraceae	Iberian star-thistle		-	+	-	-
7	*Cichorium intybus* L.	Asteraceae	Chicory		-	+	-	-
8	*Cirsium Arvense* (L.) Scop.	Asteraceae	Creeping Thistle		-	+	+	+
9	*Conyza Canadensis* (L.) Cronquist	Asteraceae	Canadian horse weed	Shaallot	+	+	-	-
10	*Conyza stricta* Wall.	Asteraceae	Horse weed/Butter weed		-	+	-	-
11	*Cynodon dactylon* (L.) Pers.	Poaceae	Bermuda grass	Dramun	+	+	+	+
12	*Dactylis glomerata* L.	Poaceae	Cock's foot		+	-	-	+
13	*Epibolium hirsutum* L.	Onagraceae	Great hairy willowherb		+	-	+	+
14	*Evolvulus nummularis* (L.) L.	Convolvulaceae	Roundleaf Bindweed		+	-	-	-
15	*Fragera vesca* L.	Rosaceae	Wild Strawberry	Ishtaber	-	+	+	+
16	*Geranium pratense* L.	Geraniaceae	Meadow geranium		-	-	-	+
17	*Geranium wallichianum* D.Don ex Sweet	Geraniaceae	Himalayan crane's bill	Ringresh/Gul e snobar	+	-	-	-
18	*Lespedeza cuneata* (Dum.Cours.) G.Don	Fabaceae	Chinese bushclober		+	-	-	-

S.no	Species	Family	Common name	Local name	Tangmarg	Khaipora	Ferozpora	Baderkut
19	*Oxalis corniculata* L.	Oxalidaceae	Creeping woodsorrel		-	-	-	+
20	*Phytolacca acinosa* Roxb.	Phytolaccaceae	Indian Poke wood	Hapat makai	-	+	+	-
21	*Poa balbosa* L.	Poaceae	Meadow grass		-	-	+	-
22	*Prunella vulgaris* L.	Lamiacea	Common self-heal	Kal wauth	-	-	-	+
23	*Ranunculus arvensis* L.	Ranunculaceae	Buttercup	Chirem	-	-	+	-
24	*Ranunculus Hirtellus* Royle	Ranunculaceae	Softly hairy buttercup		+	-	-	-
25	*Rumex acetosa* L.	Polygonaceae	Gardner sorrel	Abij	+	-	-	+
26	*Rumex hastatus* L.	Polygonaceae	Curled sock	Jungli abij	+	-	+	+
27	*Setaria viridis* (L.) P.Beauv	Poaceae	Green Fox tail	Shaal Lout	+	+	+	+
28	*Stipa sibirica* (L.) Lam.	Poaceae	Needle grass		+	+	-	+
29	*Taraxacum officinale* F.H.Wigg.	Asteraceae	Dandelion	Hund	+	+	+	+
30	*Trifolium pratense* L.	Fabaceae	Red clover	Batakh neur	+	-	-	+
31	*Trifolium repens* L.	Fabaceae	White clover	Batakh neur	+	+	+	-
32	*Utrica dioica* L.	Utricaceae	Stinging needle	Soi	+	+	+	+
33	*Verbasicum thapus* L.	Scrophulariaceae	Mullein	Wantamookh	-	+	-	-
34	*Viola odorata* L.	Violaceae	Sweet violet	Banafsha	+	-	-	+
	Fern							
	Adiantum venustum	Pteridaceae	Evergreen maidenhair		-	-	-	+

Where,
+ = presence of species
- = absence of species

Shanon diversity index

To show the species richness, Shannon's diversity was calculated from IVI values. The data present in Table 10 revealed that maximum Shannon diversity index (0.81) among trees at different sites was recorded at Tangmarg while as it was recorded minimum (0.47) at Ferozpora. Among shrubs present at different sites maximum diversity index (0.72) was obtained at Baderkut however it was minimum (0.52) at Khaipora block. The maximum diversity index (2.52) for herbs was recorded at Baderkut and the minimum diversity index (0.97) was recorded for Ferozpora block respectively.

Table 3: Shanon diversity index of Trees,Shrubs and herbs at different blocks

	Khaipora	Tangmarg	Ferozpora	Baderkut
Trees	0.54	0.81	0.47	0.52
Shrubs	0.52	0.64	0.65	0.72
Herbs	1.03	1.09	0.97	2.52

Discussion

The present study was aimed to assess the "phytosociology of Gulmarg Forest Range of Kashmir Himalayas". The results emerged have been discussed to establishing cause and effect relationship as:

1 Species composition of different conifer stands of Gulmarg forest range.

2 Phytosociology of Gulmarg Forest Range

Species composition of different conifer stands of Gulmarg forest range

The results recorded pertaining to species composition is interpreted and related to the literature to understand the Composition of species. The present study conducted at different blocks of Gulmarg Forest Range revealed that a total fifty four plant species belonging twenty five families were present. A total of eight tree species, eleven shrub species, thirty four herb species and one fern species were recorded in Gulmarg forest range (Table 2). Out of the total families recorded, Asteraceae was the dominant family represented by eight species followed by Poaceae with six species. Rosaceae and Pinaceae with four species in each family. Geraniaceae, Adoxaceae, Ranunculaceae and Polygenaceae with two species in each. While as Sapindaceae, Juglandaceae, Taxaceae, Hamamelidaceae, Apiaceae, Saxifragaceae, Onagraceae, Oxalidaceae, Phytolaccaceae, Lamiaceae, Utricaceae, Scrophulariceae, Violaceae and Pteridaceae were represented by only one species in each.

The Species composition revealed that the occurrence of similar species at the studied area may be due to the reason that all species in a natural community are not isolated. They have exhibited a positive correlation among themselves because of the interactions between the species or of the responses of the species to the same environment (Table 2). These studies had indicated the spatial distribution pattern of one species to another in a community (Misra and Misra, 1981). Similar results were also reported by Singh *et al.* (2019) Dar *et al.* (2014), Bisht *et al.* (2019) and Mir *et al.* (2019), depicting the dominance of Asteraceae family while working on different forest stands of Jammu and Kashmir Himalayas. However, Mughloo *et al.* (2021), Rizvi (2020) and Khuroo *et al.* (2006) recorded the dominance of Poaceae family followed by Asteraceae while working in similar forests of the region. While as, Bhat (2018); Gaston *et al.* (2008) reported higher number of species while assessing the floristic diversity of different protected areas of Himalayas. The Phytosociological studies revealed that the total number of tree species decreased with increase in altitude at Gulmarg forest range. According to Whittaker (1972) the altitudinal gradient is an important factor affecting species composition and structure. A gradual monotonic decrease in species richness with increasing altitude is considered as a general pattern (Francis and Currie, 1998). The decrease in number from lower elevation to higher elevation might probably be due to variation in edaphic and climatic conditions of lower temperature and short growing season at upper elevation. The results are in accordance with the findings of number of researcher like Gatoo *et al.* (2021), Rizvi (2021), Ali (2011), Dar (2018), Lanker (2007), Sharma (2006), Chandra *et al.* (1999) and Joshi (1986) who worked on different forest stands in North-West Himalayas.

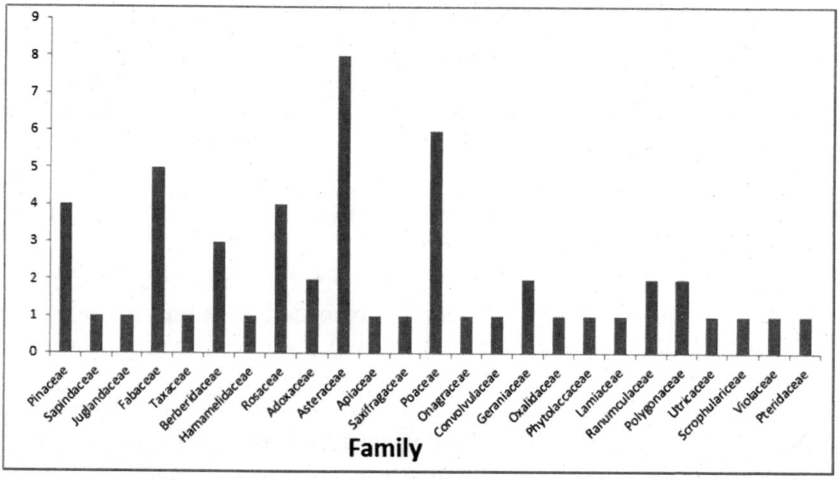

Fig. 4: Family wise species composition

Shannon diversity index

The value of diversity index of trees was 0.47-0.81 (Table 3) which is in agreement with the values (1.86-2.73) reported by Ghildiyal *et al.* (1998) (0.70-3.08) Uniyal *et al.* (2010) (0.78-3.45) Raturi (2012) (0.74-2.66) (Pant and Sammant, 2012) (0.66-2.69) Singh *et al.* (2014) (1.1-2.05) Verma (2016). Shaheen *et al.* (2012) noted its value between 0.75 and 2.27 while studying species composition and community structure of Western Himalayas from moist temperate forests of Pakistan Himalaya. The diversity index of shrubs was 0.52- 0.72 which is within the range of 1.05-2.57 as reported by Gairola *et al.* (2008) in sub-alpine zone of Western Himalaya, India and 1.7-2.5 as reported by Verma (2016) at District Chamba, Himachal Pradesh. The index of herbs in our study ranged between 0.97-2.58 which is well within the range of 2.40-3.35 as reported by Gairola *et al.* (2008) in sub-alpine zone of Western Himalaya, India; 2.49-3.01 as reported by Bharali *et al.* (2011) in west siang district of Arunachal Pradesh, India and 3.16-3.20 as reported by Verma, 2016) at District Chamba, Himachal Pradesh. The diversity index of shrubs and herbs at three blocks of Gulmarg forest range showed a decreasing trend from lower altitude to upper altitude. This is in agreement with the trend of diversity index of shrubs and herbs as has been reported by Verma and Kapoor (2011).

Conclusion

- The species composition of the study area manifested the presence of 54 plant species which include 8 trees, 11 shrubs, 34 herbs and 01 fern.
- Highest number of species (08) was recorded from Asteraceae family which was followed by Poaceae with 06 species.
- The maximum Shannon diversity was recoded at Tangmarg for trees, while as at Baderkut for shrubs and herbs.

References

Ahmed, M. H., Sheikh, T., Hussain, A. H. and Siddique, S. S. 2006. Phytosociology and structure of Himalayan forests from different climatic zones of Pakistan. *Pakistan Jouranl of Botany* **38**(2): 361-383.

Amjad, M. S., Arshad, M. and Qamar, I. A. 2013. Phytosociology of *Pinus-qurecus* forest vegetation of Nikyal hills, district Kotli, Azad Kashmir, Pakistan. *International Journal of Agriculture and Crop Sciences* **24**(5): 2952-2960.

Awan, S. M., Shaheen, D. H., Khan, R. W. A., Aziz, S. and Habib, T. 2021. Diversity and distribution pattern of Alpine vegetation communities from Ratti Gali Lake and its adjacent areas, Kashmir Himalayas, Pakistan. *Pakistan Journal of Biological sciences* **53**(2): 665-672.

Bharali, S., Paul, A., Khan, L. M. and Singha, B. L. 2011. Species diversity and community structure of a temperate mixed Rhododendron forest along an altitudinal gradient in west Siang district of Arunachal Pradesh, India. *Nature and Science* **9**(12): 101-105.

Bhat, A. A. 2017. Phytosociological status and assessment of natural regeneration of conifers in Batkote block of Pahalgam range. M. Sc. *Thesis*, Sheri Kashmir University of Agricultural Science, Kashmir.

Bhat, F., Mahajan, D. M., Sayyed, M. R. G. and Bhat, A. 2015. Phytosociological measurments and vegetation analysis along an altitudinal gradient of Lolab valley Kashmir Himalaya, India. *WCS's National Journal of Interdisciplinary Research* **1**: 1-13.

Curtis, J.T. and McIntosh, R.P. 1950. The interrelations of certain analytic and synthetic phytosociological characters. Ecology, 31(1): 434-455.

Curtis, J.T. 1959. The vegetation of winconsion on ordination of plant communities. University of Winconsion Press, Madison, Wasconsion.

Dar, G. H. and Khuroo, A. A. 2013. Floristic diversity in the Kashmir Himalaya: Progress, problems and prospects *Sains Malaysiana* **42**(10): 1377-1386.

Geelani, S. N., Bhat, G. N., Mughal, A., Husain, M., Qaiser, Q. N. and Khan, P. A. 2018. Phytosociological status of trees and shrubs of Romshi range of Shopian forest division in Kashmir valley. *Journal of Pharmacognosy and Phytochemistry* 7(3): 696-702.

Hussain, M. S., Sultana, A., Khan, J. A. and Khan, A. 2008 Species composition and community structure of forest stands in Kumaon Himalaya, Uttarakhand, India. *Tropical Ecology* **49**(2): 167-181.

Khan, A., Ahmed, M., Siddiqui, M. F., Iqbal, J. and Wahab, M. 2016. Phytosociological analysis of pine forest at Indus Kohistan, kpk, Pakistan. *Pakistan Journal of Bot*any **48**(2): 575-580.

Lan, S. 2003. Plant species diversity in Wuyishan national nature reserve. *Scientia Silvae Sinicae* **29**(1): 36–43.

Manzoor, J. and Jazib, M. J. 2020. Distribution and Phytosociological study of Agroforestry trees in Poonch district of Jammu and Kashmir. *Journal of Biodiversity Conservation and Bioresource Managmaent* **6**(2): 77-82.

Mir, N. A., Masoodi, T. H., Geelani, S. M., Wani, A. A., Parrey, G. N. and Mugloo, J. A. 2019. Floristic diversity along altitudinal gradient under Betula utilis in North Western Himalayas of Kashmir, India. *Acta Ecologica Sinica* **39**(5): 362-371.

Misra, M. K. and Misra, B. N. 1981. Species diversity and dominance in a tropical grassland community, Folia Geobot. *Et Phytotaxo* **24**(16): 309-316.

Mugloo, J. A, Adil, M., Pala, A. N, Rashid, M. 2021. Floristic Diversity along Altitudinal Gradient in Shopian Forest Range of J. & K., India. *The Indian Forester* **147**(5): 449-454.

Nafeesa, Z, and Rashid, N. 2021. Floristic, life-form, leaf-size spectra and habitat diversity of vegetation in Bhimber hills of Kashmir Himalayas. *Science Direct* **41**(3): 228-234.

Palit, D., Pal, S. and Chanda, S. 2012. Diversity and richness of plants in Darjeeling Himalaya with an eye on Gaddikhana forest beat, Senchal East zone forest range, Darjeeling. *Indian Journal of Forestry* **35**: 39-44.

Pandey, K. P., Adhikari, Y. P. and Weber, M. 2016. Structure, composition and diversity of forest along the altitudinal gradient in the Himalayas, Nepal. *Applied Ecology and Environmental Research* **14**(2): 235-251.

Philips, E.A. 1959. Methods of vegetation study. A Holt-Dryden book, Henry Holt and Co. Inc. New York.

Chapter 4

The numerous medicinal properties and phytochemistry of Hawthorn, *Crataegus* species

***Jauhar Rafeeq**[*1]**, K.N Qaisar**[2]**, Oyais Ahmad Wagay**[1]**, Mir Adil**[3]**, and Mohammad Kaif**[4]*

Division of Silviculture and Agroforestry, Faculty of Forestry SKUAST-K[1]
Controller Examination SKUAST-K[2]
Division of Forest Biology and Tree Improvement, Faculty of Forestry SKUAST-K[3]
Division of Natural Resource Management, Faculty of Forestry SKUAST-K[4]

Email: jauharparrey93@gmail.com

Abstract

Hawthorn, *Crataegus* is a genus of over 1000 species, belonging to the subfamily Maloideae in the Rosaceae family and is widely distributed in Asia, Europe and North America. Hawthorn, *Crataegus* is an important wild edible species having various medicinal benefits. It is reported that various parts of hawthorn such as fruits and leaves have several medicinal properties, which are mostly used as antispasmodic, cardiotonic, hypotensive, diuretic and atherosclerotic agents. *Crataegus* fruits contain high amounts of phenolic compounds, which are used as medicinal remedies with a variety of biological activities. *Crataegus* fruits and leaves are excellent source of antioxidants, due to the highly rich phenolic compositions and some well-known antioxidant compounds namely, hyperoside, isoquercetin, epicatechin, chlorogenic acid, quercetin, rutin, and protocatechuic acids. Further evaluation is needed to expose fully the medicinal value of the plant in order to combat various health-related complications.

Keywords: *Crataegus*, Hawthorn, phenolic compounds, Rosaceae, medicinal benefits

Introduction

Wild edible plants such as hawthorn have been an indispensable part of human life for ages. Almost every part of these plants including, their fruits, seeds, leaves, flowers, even roots and branches have been used to meet personal as well as social needs such as serving food, curing diseases and beautifying the planet (Hricova *et al.* 2016). *Crataegus* is a genus of over 1000 species, belongs to the subfamily Maloideae in the Rosaceae family. This species has a wide distribution in Asia, Europe and North America (Alirezalu *et al.* 2018). *Crataegus* is locally known by different vernacular names. In Kashmir valley it is known as Ring kul, Ban-Sangli (Hindi and Punjabi), Goni (Chitrali), Dakh (Kishtwar), Batsinga (Allai Valley and Western Himalaya Pakistan), Pingyat (Lahaul Himachal Pradesh) and Dolana (Guldara, Kabul) (Anwar *et al.* 1979; Shah and Hussain 2012; Kumar *et al.* 2009; Haq, 2012; Rawat *et al.* 2010).

It is reported that various parts of hawthorn such as fruits (Fig 1), leaves (Fig 2) and flowers (Fig 3) have several medicinal properties, which are mostly used as cardiotonic, antispasmodic, hypotensive, diuretic and atherosclerotic agents (Nabavi *et al.* 2015). Its fruits contain are rich in phenolic compounds, which are used as medicinal remedies with a variety of biological activities (Li and Wang, 2011). Besides phenolic compounds, fruits and leaves are also excellent source of well-known antioxidant compounds namely, hyperoside, isoquercetin, epicatechin, chlorogenic acid, quercetin, rutin, and protocatechuic acids. (Zugic *et al.* 2014; Barros *et al.* 2010). This paper, intends to present a review of phytochemical composition and medicinal benefits of Crataegus species.

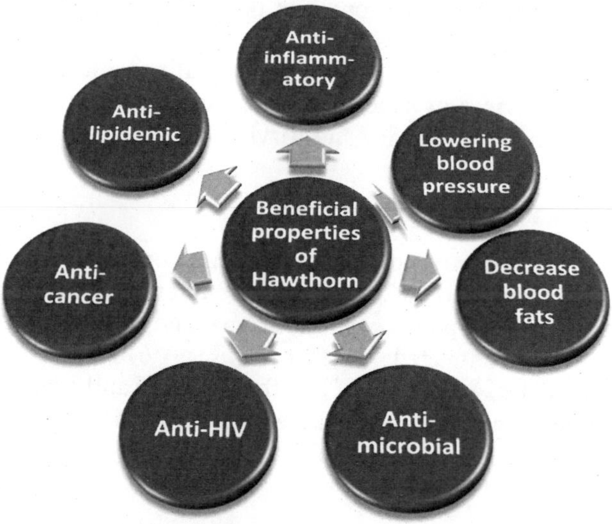

Fig 4: Beneficial properties of Hawthorn

Distribution

Crataegus is widely distributed in Afghanistan, Iran, northern India, northern Pakistan, Tadzhikistan, Kyrgyzstan, Uzbekistan, Kazakhstan, and Sinkiang at an altitudinal range of 800-2700 m. In India, it is widely found in river valleys and ravine slopes in the mountain zone and grows well in the temperate Himalayas of Kashmir and Himachal Pradesh at an altitude of 1800-3000 m (Nadkarni, 1976). It is also found in Afghanistan and Uttar Pradesh at an altitude of 1500-2700 m. In Jammu and Kashmir, it is distributed in Lolab, Sindh valley, Gulmarg, Pahalgam and Pirpanjal Range (Lone *et al.* 2014). *Crataegus songarica* is widely found in Kashmir Valley at an altitude of 1700-1900 meters above sea level (Stewart, 1972).

Taxonomical classification

Kingdom	: Plantae
Class	: Dicotyledoneae
Order	: Rosales
Family	: Rosaceae
Sub family	: Maloideae
Genus	: *Crataegus*

Medicinal properties of Hawthorn

Fruits of hawthorn are rich source of polyphenols, powerful antioxidant constituents found in plants (Zao *et al.* 2017). Presence of polyphenols in hawthorn are associated with various health promoting benefits such as lowering the risks of cancer, diabetes, heart problems, asthama and premature skin aging (Pandey and Rizvi, 2009; Wu *et al.* 2017).

Anti-inflammatory properties

Many researchers have reported anti-inflammatory properties of hawthorn (Pandey and Rizvi, 2009). Chronic inflammation has been linked to many diseases, including type-2 diabetes, asthma, and certain cancers (Hunter, 2012). A study conducted by (Han *et al.* 2016) showed that in mice having liver disease, extracts of hawthorn significantly reduced levels of inflammatory compounds. In a mice with asthama, hawthorn fruit extracts decreased inflammation and significantly reduced symptoms of asthama (Shin *et al.* 2012).

Lowering blood pressure

In traditional Chinese medicine, hawthorn fruit is one of the most commonly used food for treating blood pressure (Zou, 2016). Several studies conducted on animals show that hawthorn has the ability of lowering blood pressure (Kim *et al.* 2000). A 10 week study conducted by Walker *et al.* (2002) on 36 people with blood pressure showed that, those patients taking 500 mg of hawthorn on daily basis experienced no significant decrease in blood pressure, however a trend towards reduced diastolic blood pressure was seen in patients. In another study conducted by (Walker *et al.* 2006) showed that people with type 2 diabetes and high blood pressure, 1200 mg of hawthorn extract daily improves blood pressure and diabetes concerns in patients.

Decrease blood fats

Hawthorn has tendency of decreasing blood fats. In a mice, giving two doses of hawthorn extract lowered cholestrol, as well as triglyceride levels, compared with those mices that were not given the extract doses (Yoo *et al.* 2016). In another study conducted by (Xu *et al.* 2009), in a mice, hawthorn extract reduced total cholestrol and triglycerides contents. In another study in 264 people with anxiety, a combination of hawthorn, magnesium, and California poppy flower significantly reduced anxiety levels, compared to a placebo (Hanus *et al.* 2004).

Anti-microbial properties

Several studies showed the antimicrobial properties of hawthorn. Studies conducted by Barros *et al.* (2012) reported that the extract of hawthorn fruits showed moderate bactericidal activity against gram positive bacteria. Anti-microbial activity of *Crataegus monogyna* were studied by Belkihir *et al.* (2013). They observed that leaf and fruit extracts showed antiradical activity and anti bacterial activity against Gram positive bacteria such as *Streptococcus faecalis* and *Staphylococcus faecalis*. Studies conducted by Bisignano, (2016) showed that Apigenin-7-O-glucoside and luteolin 3,7-diglucoside extracted from hawthorn were eliminated Ureaplasma urealyticum with minimum inhibitory concentration value ranges of 0.48–3.9 μg/mL and 0.48–1.95 μg/mL, respectively.

Anti-HIV properties

Human immunodeficiency virus (HIV) releases itself from an HIV-infected cell using serine protease, followed by attack on other cells. *Crataegus pinnatifida* was found to inhibit the activity of serine protease, followed by decreasing diffusion rate of HIV in vivo (Wang *et al.* 2009). Maslinic acid isolated from *Crataegus pinnatifida* had a prominent effect on inhibiting the activity of HIV-1 protease. When the concentration was 17.9 mg/mL, the inhibitive rate is 100% (Xu *et al.* 1996). Therefore, maslinic acid is considered as a potential candidate for novel anti-HIV therapeutics.

The flavonoids and trimeric procyanidin isolated from *Crataegus sinaica* demonstrated antiviral activity against HIV. Hyperoside, vitexin, 2”-O-rhamnosylvitexin, (4’”-O-acetyl)-2”-O-rhamnosylvitexin, epicatechin, (+)-taxifolin, 3-O-β-xylopyranosyl-(+)taxifolin may exert their antiviral effects by binding to the protein coat of the virus itself or by inhibiting reverse transcriptase in retroviruses like HIV (Shahat *et al.* 1998).

Anti-cancer

The anticancer effect of maslinic acid was investigated in 1989. Maslinic acid isolated from *Crataegus pinnatifida* exhibited cytotoxicity on P-38 cancer cells (ED50 = 13.0 μg/mL) (Numata *et al.* 1989). Total flavonoids isolated from *Crataegus pinnatifida* had no effect on normal cells, but could obviously enhance calcium concentration in tumor cells. In vitro, total flavonoids inhibited and killed Hep-2 tumor cells by calcium overload, as well as inhibited DNA biosynthesis of tumor cells (Zhang *et al.* 2004). Aqueous extracts of *Crataegus pinnatifida* were found to inhibit sperm distortion of mice induced by cyclophosphamide, which obviously reduced the number of distorted sperm. The mechanism was related to the abundant linoleic acid and vitamin C in *Crataegus. pinnatifida* (Cui *et al.* 2002).

The administration of homogeneous polysaccharide (HPS) extracted from hawthorn at a concentration of 125–1000μg/mL showed anticancer activity against a human colon cancer cell line HCT116, after12h by arresting the cell cycle and inducing cell apoptosis through extrinsic and intrinsic mechanisms using P38 mitogen-activated protein kinase and the phosphatidylinositol-3-kinase/AKT/mammalian target of rapamycin signaling pathway (Ma *et al.* 2020). Quercitin 3-O-galactoside and kaempherol-3-O-glucoside inhibited the culture of MCF-7 human breast cancer cells. (Mraihi *et al.* 2015). Pinnatifidanin BVI extracted from hawthorn had a preventive effect against Mrc5 human lung cells (Huang *et al.* 2013).

Anti-lipidemic

The treatment of high-fat-diet-fed rats with L. plantarum grade A pasteurized milk ordinance -fermented hawthorn juice for 28 days showed hypolipidemic activity through the regulation of adipose tissues and liver morphology, the restoration of liver tissue and the reduction in low-density lipoprotein cholesterol, serum total cholesterol, lipid vacuolization and lipid metabolism levels (Gan, 2019).

Investigations have shown that the main antihyperlipidemic effect constituents of *Crataegus pinnatifida* are hyperin and ursolic acid. Two animal models of hyperlipidemia were established in mice with 75% yolk and Triton-WR 1339 400 mg/kg (ip), respectively. The animals were administrated with hyperoside or ursolic acid extracted from C. pinnatifida in two doses. The total cholestrol (TCH), triglyceride (TG), high density lipoprotein (HDL) and superoxide dismutase (SOD) activities in serum were measured. In comparison with control groups, TCH levels in all the dosed groups were significantly decreased, while HDL and SOD activity increased; the ratio of total cholesterol/high-density lipoprotein (TC/HDL) was reduced, too. This effect could lessen damages to vascular endothelium induced by oxygen free radical (OFR) in hyperlipoidemia, thus preventing atherosclerosis (Li, 2002; Yang, 2008). Total flavonoids of *Crataegus pinnatifida* had significant antihyperlipidemic effects and enhanced the vascular function of hyperlipidemia model rats, the mechanism of which might be relevant to the increased levels of nitric oxide (NO) in the serum and the reduction in endothelin (ET) synthesis (Li, 2002; Yang, 2008). The treatment of high-fat-diet-fed rats with L. plantarum grade A pasteurized milk ordinance -fermented hawthorn juice for 28 days showed hypolipidemic activity through the regulation of adipose tissues and liver morphology, the restoration of liver tissue and the reduction in low-density lipoprotein cholesterol, serum total cholesterol, lipid vacuolization and lipid metabolism levels (Gan, 2019).

Phytochemical Composition of Hawthorn

The genus Crataegus is reported to have phytochemicals such as oligomeric procyanidins, flavonoids, triterpenes, polysaccharides, catechloamines etc. (Kumar *et al.* 2012). It is reported that various parts of hawthorn such as fruits, leaves and flowers have several medicinal properties, which are mostly used as antispasmodic, cardiotonic, hypotensive, diuretic and atherosclerotic agents (Nabavi *et al.* 2015). Crataegus fruits contain high amounts of phenolic compounds, which are used as medicinal remedies with a variety of biological activities (Li and Wang, 2011). Crataegus fruits and leaves are excellent source of antioxidants, due to the highly rich phenolic compositions and some well-known antioxidant compounds namely, hyperoside, isoquercetin, epicatechin, chlorogenic acid, quercetin, rutin, and protocatechuic acids (Zugic *et al.* 2014; Barros *et al.* 2010). Hawthorn flowers, leaves and fruits contain different types of bioflavonoid like complexes. Major biflavonoids found in *Crataegus* species (hawthorn) include Vitexin, oligomeric procyanidins (OPC), hyperoside and quercetin. Besides these biflavonoids, hawthorn also contain other chemical constituents like, saponins, vitamin C, cardiotonic amines, purine derivatives (adenosine, adenine, guanine, caffeic acid, amygdalin), ursolic acid and triterpene acids. The chemical compounds present in Crataegus species, mainly quercetin, hyperoside, rutin, and vitexin, have been also studied using HPLC-UV and UV-Visspectrophotometry (Sagaradze *et al.* 2019). The hawthorn fruit examined by spectro photometry at a wavelength of 285±2 nm revealed the presence of hyperoside flavonoid in an amount up to 0.112–0.183% (w/w) (Kurkina, 2015). Polyphenol compounds from *Crataegus oxyacantha* extracts, including epicatechin, epicatechin gallate (ECG), rutin, caffeic, and caftaric acids, were identified using HPLC-DAD and LC-MS/MS techniques (Benabderrahmane *et al.* 2018). Rutin and quercetin were extracted from *Crataegus oxyacantha* fruit extracts using HPLC (Cuevas-Durán *et al.* 2017).

Studies conducted by various researchers have reported the presence of more than 150 bioactive compounds in the hawthorn such as phenolic acids (ferulic, gallic, p-coumaric, syringic, chlorogenic, caffeic), quercetin, pyrocatechin, phlorodizin, terpenoids, lignans, steroids, organic acids (fumaric, tartaric, succinic, citric, malic), and sugars (maltose, sucrose, glucose, fructose) (Cuevas-Durán *et al.* 2017). Different reported flavanols and flavanol polymers in *Crataegus pinnatifida* include (+)-catechin, (−)-Epicatechin, leucocyanidin ,proanthocyanidin A2, procyanidin B2, B4, B5, and trimers include procyanidin C1, procyanidin D1, epicatechin-(4β-6)-epicatechin-(4β-8)-epicatechin, epicatechin-(4β-8)-epicatechin-(4β-6)-epicatechin and procyanidin E1 (Wu *et al.* 2014). Alirezalu *et al.* (2018) reported that the total amount of phenolics ranged from 7.21 to 87.73 mg GAE/gin dry weight of the plant leaves of Crataegus species while as the total amount of flavonoids varied among species as well as in different organs of the plant (2.27 to 17.40 mg/g dry weight). Vitexin, chlorogenic acid and vitexin2"-O-rhamnoside were found to be the most abundant phenolic compounds in the extracts of Crataegus leaves (Alirezalu *et al.* 2018). Alirezalu *et al.* (2020) reported that the total phenols, total flavonoid content, and antioxidant activity were in the range of 21.19–69.12 mg gallic acid equivalent (GAE)/g dry weight (dw), 2.44–6.08 mg quercetin equivalent (QUE)/g dw and 0.32–1.84 mmol Fe++/g dw, respectively. Hyperoside (0.87–2.94 mg/g dw), chlorogenic acid (0.06–1.16 mg/g dw), and iso-quercetin (0.24–1.59 mg/g dw) were found to be the most abundant phenolic compounds in the extracts of crataegus fruits.

The antioxidant and pharmacological effects of *Crataegus* species is mainly attributed to the polyphenolic contents (Keser *et al.* 2012). Liu *et al.* (2011) applied HPLC-UV/ESI-MS to determine the phenolic constituents of hawthorn, which was found to contain C-glycosyl flavones, hyperoside, procyanidins B2/C1, and epicatechin. Lund *et al.* (2020), by using nuclear magnetic resonance (NMR) spectrometry, identified chlorogenic acid and flavonoids of Crataegus species, including vitexin-200-O-rhamnoside, rutin, hyperoside, and naringenin. Apart from these bioactive compounds, glucose, mannose and fructose were also extracted from hawthorn fruits by acid hydrolysis using 2 M trifluoroacetic acid, and then identified and characterized by gas chromatography/ mass spectrometry (Sydora *et al.* 2018). Table 1 reports the main compounds found in hawthorn.

Table 1: Main compounds found in hawthorn

Species	Compounds Identified	References
Crataegus oxyacantha	Naringenin, epicatechin, quercetin-3-O-β-glucoside, and quercetin	Benabderrahmane *et al.* 2018
Crataegus oxyacantha	Rutin and Quercetin	Cuevas-Duran *et al.* 2017
Crataegus songarica	Quercitin 3-O-galactoside and kaempherol-3-O-glucoside	Mraihi *et al.* 2015
Crataegus azarolus	Quercetin 3-O-methyl ether, and 3-β-O acetyl ursolic acid	Abu-Gharbieh *et al.* 2017
Crataegus pubescens	(+)-catechin and (–)-epicatechin	Guo *et al.* 2019
Crataegus oxyacantha	Epicatechin, epicatechin gallate (ECG), rutin, cafeic and caftaric acids	Benabderrahmane *et al.* 2018
Crataegus oxyacantha	Rutin and Quercetin	Cuevas-Durán *et al.* 2017
Crataegus oxyacantha	chlorogenic acid, vitexin 2″-O-rhamnoside, vitexin, rutin, hyperoside, quercetin, and isoquercetin	Alirezalu *et al.* 2018
Crataegus oxyacantha	vitexin, rutin, hyperoside, quercetin, and isoquercetin	Alirezalu *et al.* 2020
Crataegus monogyna	b-Carotene, Ascorbic acid, quercetin, and epicatechin	Barros *et al.* 2012

Conclusion

Different parts of hawthorn, Crataegus species such as fruits, leaves and flowers have several medicinal properties, which are mostly used as antispasmodic, cardiotonic, hypotensive, diuretic and atherosclerotic agents. These medicinal values may be linked with the presence of different phytochemicals. Further research is needed to study the relationship between the chemical composition of the species and their mode of use in the treatment of various diseases.

References

Abu-Gharbieh, E., Shehab, N.G. Therapeutic potentials of *Crataegus azarolus* var. eu-azarolus Maire leaves and its isolated compounds. BMC Complement. 2017. *Alternative Med*icine, **17**: 218- 240.

Alirezalu, A., Ahmadi, N., Salehi, P., Sonboli, A., Alirezalu, K,; Mousavi Khaneghah, A., Barba, F.J., Munekata, P.E. and Lorenzo, J.M. 2020. Physicochemical Characterization, Antioxidant Activity, and Phenolic Compounds of Hawthorn (Crataegus spp.) Fruits Species for Potential Use in Food Applications. *Foods*, **9**: 4-36.

Alirezalu, A., Salehi, P., Ahmadi, N., Sonboli, A., Aceto, S., Maleki, H.H. and Ayyari, M. 2018. Flavonoids profile and antioxidant activity in flowers and leaves of hawthorn species (*Crataegus* spp.) from different regions of Iran. *International Journal of Food Properties*, **21**(1): 452-470.

Anwar, A.K., Ashfaq, M. and Ali, M.N. 1979. Pharmacognostic studies of selected plants of Pakistan. *Forest Institute, Peshawar*, **6** (7): 114-117.

Barros, L., Carvalho, A.M. and Ferreira, I.C.F.R. 2010. Comparing the composition and bioactivity of *Crataegus monogyna* flowers and fruits used in folk medicine. *Phytochemical Analysis*, **22**: 181-188.

Barros, L., Dueñas, M., Carvalho, A.M., Ferreira, I.C. and Santos-Buelga, C. 2012. Characterization of phenolic compounds in flowers of wild medicinal plants from Northeastern Portugal. *Food and Chemical Toxicology*, **50**: 1576-1582.

Belkhir, M., Rebai, O., Dhaouadi, K., Congiu, F., Tuberoso, C.I., Amri, M. and Fattouch, S. 2013. Comparative analysis of Tunisian wild Crataegus azarolus (yellow azarole) and *Crataegus monogyna* (redazarole) leaf, fruit, and traditionally derived syrup: Phenolic profiles and antioxidant and antimicrobial activities of the aqueous-acetone extracts. *Journal of Agricultural and Food Chemistry*, **61**: 9594-9601.

Benabderrahmane, W., Lores, M., Lamas, J.P. and Benayache, S. 2018. Matrix solid-phase dispersion as a tool for phytochemical and bioactivities characterisation: *Crataegus oxyacantha* L. A case study. *Natural Product Research.*, **32**: 1220-1223.

Bisignano, C., Furneri, P.M. and Mandalari, G. 2016. In Vitro Efficacy of *Crataegus oxycantha* L. (Hawthorn) and Its Major Components against ATCC and Clinical Strains of Ureaplasma urealyticum. *Advanced Microbiology*, **6**: 909-916.

Cuevas-Durán, R.E., Medrano-Rodríguez, J.C., Sánchez-Aguilar, M., Soria-Castro, E., Rubio-Ruíz, M.E., Valle-Mondragón, D., Sánchez-Mendoza, A., Torres-Narvaéz, J.C., Pastelín-Hernández, G. and Ibarra-Lara, L. 2017. Extracts of *Crataegus oxyacantha* and Rosmarinus officinalis attenuate ischemic myocardial damage by decreasing oxidative stress and regulating the production of cardiac vasoactive agents. *International Journal of Molecular Sciences*, **18**: 12-24.

Cuevas-Durán, R.E., Medrano-Rodríguez, J.C., Sánchez-Aguilar, M., Soria-Castro, E., Rubio-Ruíz, M.E., Valle-Mondragón, D., Sánchez-Mendoza, A., Torres-Narvaéz, J.C., Pastelín-Hernández, G. and Ibarra-Lara, L. 2017. Extracts of *Crataegus oxyacantha* and *Rosmarinus officinalis* attenuate ischemic myocardial damage by decreasing oxidative stress and regulating the production of cardiac vasoactive agents. *International Journal of Molecular Sciences*, 18: 12-24.

Cui, T.C., Liu, X.Q., Xu, H.Q., Wu, G.J. and Zhang, Z.B. 2002. Effect of the extracts of *Crataegus pinnatifida* on inhibiting sperm distortion of mice induced by cyclophosphamide. *Chinese Journal of Public Health*, **3**: 266-267.

Guo, R., Lv, T.M., Han, F.Y., Lin, B., Yao, G.D., Wang, X.B., Huang, X.X. and Song, S.J. 2019. Chiral resolution and neuroprotective activities of enantiomeric dihydro benzo furanne olignans from the fruitof *Crataegus pinnatifida*. *Bioorganic Chemistry*, **85**: 469-474.

Han, Xiao., Li, W., Huang, D. and Yang, X. 2016. Polyphenols from hawthorn peels and fleshes differently mitigate dyslipidemia, inflammation and oxidative stress in association with modulation of liver injury in high fructose diet-fed mice. *Chemico-Biological Interactions*, **257**: 132-140.

Hanus, M., Lafon, J. and Mathieu, M. 2004. Double blind, randomized, placebo-controlled study to evaluate the efficacy and safety of a fixed combination containing two plant extracts (*Crataegus oxyacantha* and *Eschscholtzia californica*) and magnesium in mild to moderate anxiety disorders. *Current Medical Research and Opinion*, **20**(1): 63-71.

Haq, F. 2012. The Ethno Botanical Uses of Medicinal Plants of Allai Valley, Western Himalaya Pakistan, *International Journal of Plant Research*, **2**(1): 21-34.

Hricova, A., Fejer, J., Libiakova, G., Szabova, M., Gazo, J. and Gajdosova, A. 2016. Characterization of phenotypic and nutritional properties of valuable *Amaranthus cruentus* L. mutants.,*Turkish Journal of Agriculture and Forestry*, **40**: 761-771.

Huang, X.X., Zhou, C.C., Li, L.Z., Li, F.F., Lou, L.L., Li, D.M., Ikejima, T., Peng, Y. and Song, S.J 2013. . The cytotoxicity of 8-O-4' neolignans from the seeds of *Crataegus pinnatifida*. *Bioorganic and Medical Chemistry Letters* , **23**: 5599-5604.

Hunter, P. 2012. The inflammation theory disease. The growing realization that chronic inflammation is crucial in many diseases opens new avenues for treatment. *European Molecular Biology Organization Reports*, **13**(11): 968-970.

Keser, S., Celik, S., Turkoglu, S., Yilmaz, O. and Turkoglu, I. 2012. Hydrogen peroxide radical scavenging and total antioxidant activity of hawthorn. *Chem.istry Journal*, **2**: 9-12.

Kim, S.H., Kang, K.W., Kim, K.W. and Kim, N.D. 2000. Procyanidins in Crataegus extract evoke endothelium- dependent vasorelaxation in rat aorta. *Life Sciences*, **67**(2): 121-131.

Kumar, M., Paul, Y. and Anand, V.K. 2009. An Ethnobotanical Study of Medicinal Plants used by the Locals in Kishtwar, Jammu and Kashmir, India. *Ethnobotanical Leaflets*, **13**: 1240-56.

Kurkina, A. 2015. Determination of total flavonoids in siberian hawthorn fruit. *Pharmaceutical Chemistry Journal*. **48**: 800–803.

Li, C. and Wang, M.H. 2011. Anti-inflammatory effect of the water fraction from hawthorn fruit on LPS-stimulated RAW 264.7 cells. *Nutrition Research and Practice*, **5**(2): 101-106.

Li, G.H., Sun, J.Y. and Zhang, X.L. 2002. Experimental studies on antihyperlipidemia effects of two compositions from hawthorn in mice. *Chinese Traditional and Herbal Drugs*, **33**: 50-52.

Liu, P., Kallio, H. and Yang, B. 2011. Phenolic compounds in hawthorn (*Crataegus grayana*) fruits and leaves and changes during fruit ripening. *Journal of Agricultural and Food Chemistry*, **59**: 11141–11149

Lone, P.A., Bhardwaj, A.K., Shah, K.W. and Tabasum, S. 2014. Ethnobotanical survey of some threatened medicinal plants of Kashmir Himalaya, India. *Journal of medicinal plant research*, **8**(47): 13-73.

Lund, J.A., Brown, P.N. and Shipley, P.R. 2020. Quantification of North American and European Crataegus flavonoids by nuclear magnetic resonance spectrometry. *Fitoterapia*, **143**: 104-110.

Ma, L., Xu, G.B., Tang, X., Zhang, C., Zhao, W., Wang, J. and Chen, H. 2020. Anti-cancer potential of polysaccharide extracted from hawthorn (Crataegus) on human colon cancer cell line HCT116 via cell cycle arrest and apoptosis. *Journal of Functional Foods*, 64: 103-677.

Mraihi, F., Fadhil, H., Trabelsi-Ayadi, M. and Chérif, J.K. 2015. Chemical characterization by HPLC-DAD-ESI/MS of flavonoids from hawthorn fruits and their inhibition of human tumor growth. *Journal of New Sciences*, **15**: 840–846.

Mraihi, F., Hidalgo, M., de Pascual-Teresa, S., Trabelsi-Ayadi, M. and Chérif, J. K. 2015. Wild grown red and yellow hawthorn fruits from Tunisia as source of antioxidants. *Arab Journal of Chem*istry, **8**: 570-578.

Nabavi, S.F., Habtemariam, S., Ahmed, T., Sureda, A., Daglia, M., Sánchez, E.S. and Nabavi, S.M. 2015. Polyphenolic Composition of *Crataegus monogyna* Jacq: From Chemistry to Medical Applications. *Nutrients*, 7: 7708-7728.

Nadkarni, K.M. 1976. Indian Materia Medica, Popular Prakashan Private Limited, Bombay, Vol.I&II.

Numata, A., Yang, P., Takahashi, C., Fujiki, R., Nabae, M. and Fujita, E. 1989. Cytotoxic triterpenes from a Chinese medicine, goreishi. *Chemical and Pharmaceutical Bulletin*, **37**: 648-651.

Pandey, K.B. and Rizvi, S.I. 2009. Plant polyphenols as dietary antioxidants in human health and disease. *Oxidative Medicine and Cellular Longevity*, **2**(5): 270-278.

Rawat, Y.S., Vishvakarma, S.C.R., Oinam, S.S. and Kuniyal, J.C. 2010. Diversity, distribution and vegetation assessment in the Jahlmanal watershed in cold desert of the Lahaul valley, north-western Himalaya, India. *Forest– Biogeosciences and Forestry*, **3**: 65-71.

Sagaradze, V.A., Babaeva, E.Y., Ufimov, R.A., Trusov, N.A. and Kalenikova, E.I. 2019. Study of the variability of rutin, vitexin, hyperoside, quercetin in "Crataegi folium cum flore" of hawthorn (Crataegus L.) species from Russian flora. *Journal of Applied Research on Medicinal and Aromatic Plants*, **15**: 100-217.

Shah, S. and Hussain, M.F. 2012. Ethnomedicinal plant wealth of Mastuj valley, Hindukush range, District Chitral, Pakistan. *Journal of Medicinal Plants Research*, **6**(26): 4328-4337.

Shahat, A., Ismail, S. and Hammouda, F 1998. Anti-HIV activity of flavonoids and proanthocyanidins from *Crataegus sinaica*. *Phytomedicine*, **2**: 133-136.

Shin, I.S., Lee, M.Y., Lim, H.S., Ha, H., Seo, C.S., Kim, J.C. and Shin, H.K. 2012. An extract of *Crataegus pinnatifida* fruit attenuates airway inflammation by modulation of matrix metalloproteinase-9 in ovalbumin induced asthama. *PLoS One*, **7**(9): 1-11.

Stewart, R.R. 1972. An annotated catalogue of the vascular plants of West Pakistan and Kashmir. *Fakhri press, Karachi, Pakistan*.

Sydora, N.V., Kovalyova, A.M. and Iakovenko, V.K. 2018. The study of the carbohydrate composition of hawthorn fruits. *News Pharm*, **3**: 14-18.

Walker, A.F., Marakis, E.S., Simpson, E., Hope, J.L., Robinson, P.A., Hassanein, M. and Simpson, H.C.R. 2006. Hypotensive effects of hawthorn for patients with diabetes taking prescription drugs: a randomized controlled trial. *British Journal of General Practice*, **56**: 37-43.

Walker, A.F., Marakis, G., Morris, A.P. and Robinson, P.A. 2002. Promising hypotensive effect of hawthorn extract: a randomized double-blind pilot study of mild, essential hypertension. *Phytotherapy Research*, **16**(1): 48-54.

Wang, B., Qiu, W.W., Yu, Y.Y., Yang, F. and Tang, J. 2009. Progress of study on maslinic acid. *Chinese Bulletin of Life Sciences*, **21**: 264–269.

Wu, J., Peng, W., Qin, R. and Zhou, H. 2014. *Crataegus pinnatifida*: Chemical constituents, pharmacology, and potential applications. *Molecules*, 19: 1685-1712.

Wu, P., Li, F., Zhang, J., Yang, B., Ji, Z. and Chen, W. 2017. *BMC Complementary and Alternative Medicine*, **17**: 151-158.

Xu, H., Xu, Hou-En. and Ryan, D. 2009. A study of the comparative effects of hawthorn fruit compound and simvastatin on lowering blood lipid levels. *American Journal of Chinese Medicine*, **37**(5): 903-908.

Xu, H.X., Zeng, F.Q., Wan, M. and Sim, K.Y. 1996. Anti-HIV triterpene acids from Geum japonicum. *Journal of Natural Products*, **59**: 643-645.

Yang, Y.J., Lin, J. and Wang, C.M. 2008. Effect of FHL on intervening early hyperlipoidemia model rat. *Chinese Traditional and Herbal Drugs*, **39**: 1848-1850.

Yoo, J.H., Liu, Y. and Kim, H.S. 2016. Hawthorn fruit extract elevates expression of Nrf2/HO-1 and improves lipid profiles in Ovariectomized Rats. *Nutrients*, **8**(5): 1-16.

Zao, C.N., Meng, X., Li, Y., Li, S., Liu, Q., Tang, G.Y. and Li, H.B. 2017. Fruits for prevention and treatment of cardiovascular diseases. *Nutrients*, **9**(6): 598-626.

Zhang, Y., Li, H.W. and Sun, J.P. 2004. Extraction-Separation of total flavone from the fruit of *Crataegus pinnatifida* and anticancer activities in vitro. *Chinese Traditional and Herbal Drugs*, **35**: 787-789.

Zou, P. 2016. Traditional Chinese medicine, Food therapy, and hypertension control: A narrative review of Chinese literature. *The American Journal of Chinese Medicine*, **44**(8): 1-16.

Zugic, A., Dordevic, S., Arsic, I., Markovic, G., Zivkovic, J., Jovanovic, S. and Tadic, V. 2014. Antioxidant activity and phenolic compounds in 10 selected herbs from Vrujci Spa, Serbia. *Industrial Crops and Products*, **52**: 519-527.

Chapter 5

Floristic diversity of an urban green space (Iqbal Park) in Srinagar, Kashmir

*Muazzama Mushtaq, Anam Gowher, Suraya Tasveer, Baria Ashraf, Khurshid Ahmad Sofi and Aasif Ali Gatoo**

Division of Natural Resource Management, Faculty of Forestry SKUAST- Kashmir

**asifskuast@gmail.com*

Abstract

The present investigation entitled "Floristic diversity of an urban green space (Iqbal Park) in Srinagar, Kashmir" were carried out at Iqbal Park, Srinagar, during the Year 2021 and 2022. Extensive field survey was conducted in the study area during the year 2021. The survey was conducted throughout the year. In the first instance efforts were made to identify various species present in the study area. During the study all species including trees, shrubs, herbs and ornamental plants were taken into consideration. Help was taken from the taxonomist as a well as people who were shown samples of plants as well as pictures of plants identification. Investigations conducted revealed that the study area is having rich floristic diversity. A total of 64 species were recorded in the study area in which 27 were trees, 11 shrubs, 15 flowering plants, 10 herbs and 1 grass species. These species belong to 35 families. Among all the families, Asteraceae was recorded as dominant family in which 13 species were recorded in the study area.

Key words: *Floristic diversity, Iqbal Park, Srinagar, Trees, Shrubs, Herbs*

Introduction

The term "biodiversity" was first coined in late 1987 (Wilson 1988) as contraction of the phrase "biological diversity" proposed about 15 years earlier, though some components of the concept have been familiar since ancient times in the history of science. Biodiversity refers to the variety and variability of living organisms and of the ecological complexes in which they occur (Dar *et al.* 2002). One of its primal and easy definitions is that of Wilcox (1984), according to which biological diversity is defined as "the diversity of life forms in a given area, the ecological responsibility they perform and the genetic diversity they possess."

Floral diversity refers to the diversity of plants that occurs in a specific region during distinct era. It broadly refers to the diverseness of naturally occurring indigenous or native plants. A complete of 215,644 species of plants from a total of 298,000 predicted have been catalogued on planet so far. Alone, 8,600 plant species have been reported from ocean out of evaluated 16,600.

India, being one of the tropical countries which is located in the middle of 8°4' north to 37°6' north latitude and 68°7' east to 97°25' east longitude, with about 7500-km-long coastline and over 15,000-km-long land borderline, harbours 46,824 species including species of virus/bacteria and fungi. The floral diversity in India is found in four phytogeographically distinct regions, which include, Himalayas, Western Ghats, Northeast India and Andaman and Nicobar Islands. The Flora of India represents 11.4% of the total reported plant species of the world and around 28% species of plants are endemic to India. Angiosperms are the largest plant group in India comprising a total of 17,817 species, constitutes 38.15% of floral diversity of the entire country, followed by fungi comprised of 14,698 species, representing 31.38% (CBD, 1992). The nation as well has elevated level of cryptogam (Bryophytes and Pteridophytes) diversity. A total of 2,479 species of Pteridophytes and around 1265 of Bryophytes have been recorded from India so far. As per the recent evaluation, the Indian flora illustrates close to 12% of the global floral diversity (which excludes virus). The importance of the Flora of India is further demonstrated by the spread of species of wild relatives of crop plants in distinct regions of the country.

The UT of Jammu and Kashmir (J&K) and Ladakh, occurs between latitude 32.28-37.06° and longitude 72.53-80.32°, is situated in the north-western part of the Himalayan mountain arc in India, at a 220-8611 m altitudinal range. The vegetation of Kashmir represents significantly diverse types due to immense habitat diversity. Terrestrial and aquatic are the two main types of natural ecosystem that exist there. The major proportion of the landscapes is formed by the terrestrial ecosystem (Malik, 2020). In the UT of J&K the highest forest cover is in the Kashmir region (about 51% of its geographical area). The temperate forests lie between 1600 - 2700 m and incorporate usually the conifers, such as Blue Pine i.e *Pinus wallichiana*, Himalayan Deodar i.e *Cedrus deodara*, Silver Fir i.e *Abies pindrow*, Spruce i.e *Picea smithiana* and the Himalayan Yew i.e *Taxus wallichiana*, with some correlated broad-leaved trees and shrubs. In the subalpine forests that ranges from 2700 - 3500 m altitudinal range, the Silver Fir bears dominance in the lower reaches, while natural stands of Birch (*Betula utilis*) occur in ranges higher than 3200 m, forming the timberline in the Kashmir Himalaya. A complete of 2312 species belonging to the 'land plants' – bryophytes, pteridophytes, gymnosperms and angiosperms – have been reported from the region. In total, these species are grouped under 842 genera and 189 families (Dar *et al.* 1995). As very limited floristic studies have been conducted on the lower plant groups, such as bryophytes and pteridophytes, the number of species reported in these groups from the region is expectedly low.

Keeping in view the importance of biodiversity the present study was conducted in Iqbal Park, Srinagar, with the aim to document the Floristic diversity of the studied area.

Materials and Methods

Study Area

Present study was conducted at Iqbal Park Srinagar, which is located in the Srinagar district of Kashmir Himalayas. The study site is located at an altitude of 1585 m having coordinates 34° 3' 56" N and 74° 48' 12" E. The study area has temperate climate experiencing four distinct seasons: a severe winter (December to February), a cold spring (March to May), a mild summer (June to August) and a pleasant autumn (September to November).

Methodology

Extensive field survey was conducted in the study area during the year 2021. The survey was conducted throughout the year. In the first instance efforts were made to identify various species present in the study area. During the study all species including trees, shrubs, herbs and ornamental plants were taken into consideration Help was taken from the taxonomist as a well as people who were shown samples of plants as well as pictures of plants identification.

Experimental Finding

Investigations conducted revealed that the study area is having rich floristic diversity. A total of 64 species were recorded in the study area (Table 1) in which 27 were trees, 11 shrubs, 15 flowering plants, 10 herbs and 1 grass species (Fig 2). These species belong to 35 families. Among all the families, Asteraceae was recorded as dominant family in which 13 species were recorded in the study area.

Table 1: Floristic List of Iqbal Park, Srinagar

S.No	Species	Family	Common name	Local name
		Trees		
1	*Acacia estrophiolata*	Fabaceae	Iron wood	
2	*Acer negundo*	Sapindaceae	Box elder	
3	*Aesculus indica*	Sapindaceae	Indian-horse chestnut	Handoon
4	*Catalpa bignonioides*	Bignoniaceae	Indian bean tree	
5	*Cedrus deodara*	Pinaceae	Himalayan cedar	Deodar
6	*Crataegus songarica*	Rosaceae	Hawthorn	Ring Kul
7	*Cupressus arizonica*	Cupressaceae	Cypress	Yari kul
8	*Cupressus torulosa*	Cupressaceae	Himalayan Cypress	Yari kul
9	*Frangula alnus*	Rhamnaceae	Alder buckthorn	
10	*Ginkgo biloba*	Ginkgoaceae	Maidenhair tree	Aziz
11	*Indocalamus latifolius*	Poaceae	Indocalamus	Bhans
12	*Lagerstroemia indica*	Lythraceae	Common Crape Myrtle	
13	*Mangolia grandiflora*	Mangoliaceae	Bull bay	
14	*Nerium oleander*	Apocynaceae	Oleander	Gandula
15	*Plantinus orientalis*	Platanaceae	Chinar	Boen
16	*Platycladus orientalis*	Cupressaceae	Chinese thuja	Sarva
17	*Punica granatum*	Lythraceae	Pomeganate	Dhaaen
18	*Prunus cerasifera*	Rosaceae	Cherry plum	
19	*Prunus persica*	Rosaceae	Rosaceae	Tchenunn
20	*Pyracantha coccinea*	Rosaceae	Red firethorn	
21	*Quercus robur*	Fagaceae	English oak	
22	*Robinia pseudoacacia*	Fabaceae	Black locust	Kikar
23	*Salix babylonica*	Salicaceae	Weeping willow	Veer
24	*Salix caprea*	Salicaceae	Goat willow	Braedmushk
25	*Styphnolobium japonicum*	Fabaceae	Japanese pagoda tree	Chhatri kul
26	*Trachycarpus fortunei*	Arecaceae	Chinese windmill palm	
27	*Ulmus minor 'Jacqueline Hillier'*	Ulmaceae	Elm tree	Brenn

S.No	Species	Family	Common name	Local name
		Shrubs		
28	*Abelia X grandiflora*	Caprifoliaceae	Glossy Abelia	
29	*Berberis thunbergii*	Berberidaceae	Japanese barberry	Kawdach
30	*Buxus sempervirens*	Buxaceae	Common box	Shamshad
31	*Forsythia X intermedia*	Olives	Border Forsythia	
32	*Juniper squamata*	Cupressaceae	Flaky juniper	Yathur, Vaitro
33	*Nandina domestica*	Berberidaceae	Heavenly bamboo	Bhaans
34	*Parthenocissus quinquefolia*	Vitaceae	Virginia creeper	
35	*Peperomia*	Piperaceae	Radiator plant	
36	*Pittosporum tenuifolium*	Pittosporaceae	Kohuhu	
37	*Prinsepia utilis*	Rosaceae	Cherry prinsepia	Bhikal, Bekkra
38	*Viburnum opulus*	Adoxaceae	Water elder	
		Flowering Plants/Ornamentals		
39	*Antirrhinum majus*	Plantaginaceae/ Veronicaceae	Common snapdragon	
40	*Calendula arvensis*	Asteraceae	Field marigold	Jaffer
41	*Chrysanthemums*	Asteraceae	Mums	Gul e dawood
42	*Chrysanthemum X morifolium*	Asteraceae	Florist's daisy	Gulbahaaar
43	*Cosmos bipinnatus*	Asteraceae	Garden cosmos	
44	*Coreopsis lanceolata*	Asteraceae	Lanceleaf tickseed	
45	*Dahlia pinnata*	Asteraceae	Garden dahlia	
46	*Dianthus barbatus*	Caryophyllaceae	Sweet william	
47	*Epilobium hirsutum*	Onagraceae	Great willow herb	
48	*Gazania linearis*	Asteraceae	Treasure flower	Gazania
49	*Gomphrena globosa*	Amaranthaceae	Globe amaranth	Gundi
50	*Lycoris radiata*	Amaryllidaceae	Red spider lily	
51	*Rose*	Rosaceae	Garden rose	Gulab
52	*Rosa X damascena*	Rosaceae	Damask rose	Gulab
53	*Verbena*	Verbenaceae	Vervain	

S.No	Species	Family	Common name	Local name
		Herbs		
54	*Ageratum houstonianum*	Asteraceae	Floss flower	
55	*Celosia argentea var. cristata*	Amaranthaceae	Cockscomb	Mawal
56	*Chrysanthemum japonese*	Asteraceae	Ashizuri nojigiku	
57	*Matthiola incana*	Brassicaceae	Brompton stock	
58	*Salvia splendens*	Lamiaeae	Scarlet sage	
59	*Tagetes*	Asteraceae	Marigold	Jaffer
60	*Viola tricolor var. horthensis*	Violaceae	Pansy	Panzeen posh
61	*Zephyranthes canadida*	Asteraceae	White windflower	
62	*Zinnia angustifolia*	Asteraceae	Narrow leaf zinnia	Zinnia
63	*Zinnia elegans*	Asteraceae	Zinnia	Zinnia
64	*Cymbopogon*	Poaceae	Poaceae	

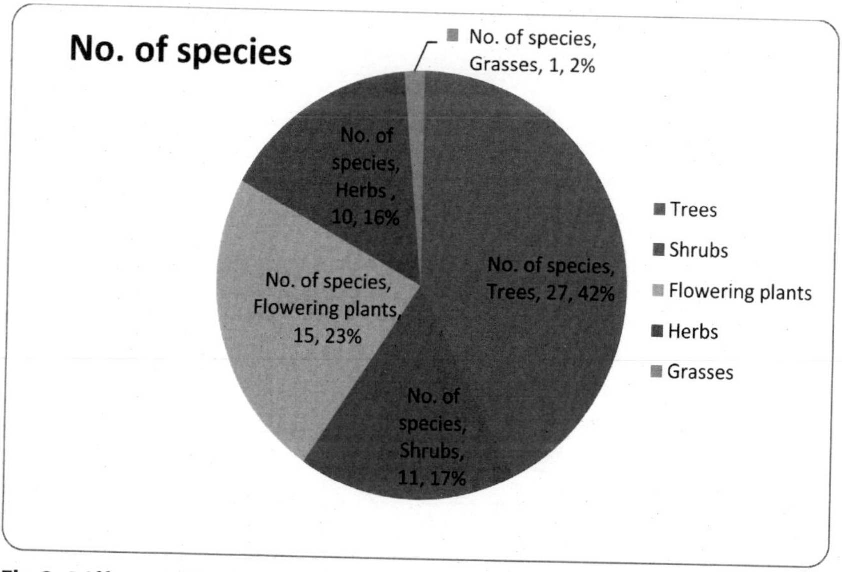

Fig 2: Different life-forms present in Iqbal Park Srinagar

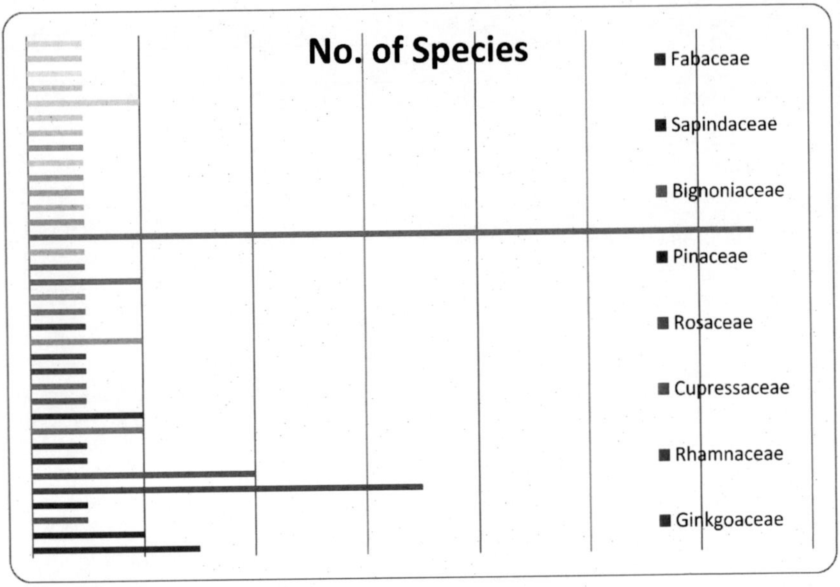

Fig 2: Different plant families in Iqbal Park, Srinagar

Table 2: Different plant families in Iqbal Park, Srinagar

S.No.	Family	No. of species
1	Fabaceae	3
2	Sapindaceae	2
3	Bignoniaceae	1
4	Pinaceae	1
5	Rosaceae	7
6	Cupressaceae	4
7	Rhamnaceae	1
8	Ginkgoaceae	1
9	Poaceae	2
10	Lythraceae	2
11	Mangoliaceae	1
12	Apocynaceae	1
13	Platanaceae	1
14	Fagaceae	1
15	Salicaceae	2
16	Arecaceae	1
17	Ulmaceae	1
18	Caprifoliaceae	1
19	Berberidaceae	2
20	Buxaceae	1

S.No.	Family	No. of species
21	Olives	1
22	Asteraceae	13
23	Vitaceae	1
24	Piperaceae	1
25	Pittosporaceae	1
26	Adoxaceae	1
27	Plantaginaceae/ Veronicaceae	1
28	Caryophyllaceae	1
29	Onagraceae	1
30	Verbenaceae	1
31	Amaranthaceae	2
32	Amaryllidaceae	1
33	Brassicaceae	1
34	Lamiaeae	1
35	Violaceae	1

Discussion

Several researchers have conducted studies on floristic diversity throughout globe. Verma (2016) reported 102 species belonging to 54 families and 95 genera in Dankund beat of Kalatop Khajjiar wild life sanctuary in district Chamba, Himachal Pradesh. The dominant families were Asteraceae, Rosaceae and Fabaceae. Khan *et al.* (2014) reported 145 threatened, 30 endangered, 68 vulnerable and 47 rare species in Poonch valley of Azad Kashmir. A total of 182 species were reported by Vijayakumari *et al.* (2019) in Palayamkottai region of Tirunelveli district of Tamil Nadu, India, belonging to 159 genera and 71 families. Amoung these dicot were dominant with 53 families and 150 species, on the other hand monocot comprised of 15 families and 29 species and gymnosperm at the third position were represented by 2 families and 2 species and a single species of Pteridophyte. Hussain (2009) reported 120 plant species belonging to 92 genera and 46 families. Poaceae was the dominating with 14 members in Bandipora valley, Kashmir Himalaya.

Conclusion

The results of the present investigation, explained with tables and figures are summarized and concluded that the study area is having rich floristic diversity. A total of 64 species were recorded in the study area.Among all the species, 27 were trees, 11 shrubs, 15 flowering plants, 10 herbs and 1 grass species. These species belong to 35 families. Among all the families,

Asteraceae was recorded as dominant family in which 13 species were recorded in the study area.

Literature Cited

C.B.D. 1992.Convention on Biological Diversity: text and annexes, see article 2. Secretariat of the Convention on Biological Diversity, Montreal, Canada, pp. 1–34.

Dar, G.H., Bhagat, R.C. and Khan, M.A. 2002. Biodiversity of the Kashmir Himalaya. Valley Book House, Srinagar, pp399.

Dar, G.H., Naqshi, A.R. & Ara, S. 1995. New records and new taxa of flowering plants from Jammu and Kashmir State, 1970-1992. *Oriental Science (Special Issue)* pp33-44.

Hussain, M. (2009). Floristic studies in bandipur valley, Kashmir Himalaya. *International Journal of Environmental Sciences* **9**(2): 81-85.

Khan, A., Ahmed, M., Siddiqui, M.F., Iqbal, J. and Wahab, M. 2016. Phytosociological analysis of pine forest at Indus Kohistan, kpk, Pakistan.*Pakistan Journal of Bot*any **48**(2): 575-580.

Malik, Z.A. 2020. Books on biodiversity and conservation. *Biodiversity and Conservation* **29**: 3860-3861.

Verma, R.K. 2016.Status of plant diversity along an altitudinal gradient in Dankund beat of Kalatop Khajjiar wild life sanctuary of district Chamba, Himachal Pradesh. *Biological Forum,* **8**(1): 540-547.

Vijayakumari, J., Prabha, V.S., Jebarubi, E., Raj, T.L.S., Antony, S.B. and Rayan. 2019. Floristic Diversity Assessment of Home Garden in Palayamkottai Region of Tirunelveli District, Tamil Nadu a Means of Sustainable Biodiversity Conservation. *International Journal of Trend in Scientific Research and Development* **3**(3): 1484-1491.

Wilcox, B.A. 1984. 'In situ' conservation of genetic resources: determinant of minimum area requirements. Proc World National Parks Congress, Bali, Indonesia. IUCN, Gland, Switzerland.

Wilson, E.O. 1988. Biodiversity. National Academy Press, Washington, DC.

Hussain, M. (2009). Floristic studies in bandipur valley, Kashmir Himalaya. *International Journal of Environmental Sciences* **9**(2): 81-85.

Chapter 6

Use of plants and animals as bio-indicator in climate change

Baria Ashraf, Shah Murtaza Mushtaq, Snowber Zehra, Suraya Tasveer, and M A Islam*

Division of Natural Resource Management, Faculty of Forestry SKUAST- Kashmir

** bariabeigh22@skuastkashmir.ac.in*

Introduction

The term 'Bioindicators' refers to any organisms, species, communities or biological processes, which help to qualitatively monitor the status of the environment in which they live by means of their function as well as their population (Pereira *et al.*, 2022). These can be plants, animals, algae, lichens, zooplanktons, insects, amphipods, molluscs, echinoderms, and other micro-organisms. However, a general, all encompassing definition of a biological indicator would be: "a species or group of species that readily reflects the abiotic or biotic state of an environment, represents the impact of environmental change on a habitat, community or ecosystem or is indicative of the diversity of a subset of taxa or the whole diversity within an area". The bioindicator has requirements about a known set of physical or chemical variables such that changes in presence/absence, numbers, morphology, physiology or behaviour of that species indicate that the given physical or chemical variables are outside their preferred limits (Oxford University Press: London, UK, 2020). The occurrence of any environmental change or problem within an ecosystem can be predicted very easily by any changes in the population status, behaviour, and physiology of such organisms (Moreno *et al.*, 2007). Bioindicators are useful in three situations: 1) where the indicated environmental factor cannot be measured, e.g., in situations where environmental factors in the past are reconstructed such as climatic change, studied in palaeo-biomonitoring 2) where the indicated factor is difficult to measure, e.g., pesticides 3) where the environmental factor is easy to measure but difficult to interprete, e.g., whether the observed changes have ecological significance. For more than 50 years, certain biological organisms have been used as bioindicators because they are sensitive to environmental factors such as climate change, ozone, or water pollution (Middleton *et al.*, 1950; Manning, 1998). Environmental bioindicators have also proved to be tools in understanding the effects of the presence of contaminants or in detecting biological changes in the environment The information provided by bioindicators is adequate and reliable and might be difficult to acquire or compute in such a quick manner using other means (Khatri and Tyagi, 2015).

The earth is adverse subject to climate change and habitat deterioration on unprecedented scales. Climate change, biodiversity loss, nitrogen and phosphorus cycles have crossed the critical levels of the threshold for the sustenance of humanity on earth (Rockstrom *et al.*, 2009). The mitigation of these factors primarily requires their early detection and thorough understanding of each component (Cash *et al.*, 2003). Traditionally, climate change monitoring involves satellite data, ground studies, and modelling of the predications (Appenzeller *et al.*, 2008). The higher maintenance cost of the traditional climate change monitoring systems and the inconsistency of monitoring data to actual events have necessitated the researchers to utilize bioindicators such as fish, insects, mussels, lichens, algae, plants, birds etc. (Guralnick 2002; Caza *et al.,* 2016). It is therefore important to identify bioindicator taxa that show measurable responses to climate change and habitat loss and that reflect wider-scale impacts on the biota of interest.

Characteristics of Bioindicators

In order to prepare adaptive strategies of biological organisms against climate change it might be the best way to assess climatic affect to all the biological taxa known. Instead of applying this impossible method to assess climatic affects to all the biodiversity, some of biological organisms are selected as biological indicators and used to provide general information about complexity between climate and biodiversity change. The most important factor in defining bioindicators sensitive to climate change is how sensitive biological organisms are to climatic elements such as temperature, precipitation, and humidity rather than other environmental factors including acidification, land use change, habitat constraints, etc. Therefore, the climate sensitive biological indicator species are defined as species which need to be consistently investigated and managed in a type of indicators because they are showing or expected to show clear change in activity, distribution or size of population owing to climate change (Rainio and Niemela, 2003). From the technical series of the Convention on Biological Diversity, Watson *et al.,* (2003) suggested that indicators must be practical and meaningful at both the national and site levels as well as consistent with the main objectives of the project or policy intervention. For indicators to be most useful and effective, they also suggested that the suite of indicators should be relevant to the project as much detailed as follows. Biological indicators should be cost effective to monitor; well established in order to reveal meaningful trends; be precise so that they can be clearly defined and understood the same way by different stakeholders; be chosen that allow the identification and separation of the effects of climate change; be evaluated by minimum number of individuals and agencies in their evaluation. In short, a good bioindicator should be well known taxonomically, be distributed over a broad geographic area, be easy and cost-effective to survey, be relatively independent of sample size, ecology and life history well understood, taxonomically well documented and stable and should be of potential economic importance (Noss, 1990; Niemela, 2000; Rainio and Niemela, 2003).

Different types of Bioindicators

(i) Microbial Bioindicator

Microbial bioindicators can be referred as the micro-organisms which are used to indicate the health of terrestrial and aquatic ecosystem. Plankton like microalgae, copepods and other small water crustaceans are the example of bioindicator species that can be monitored for biochemical, physiological and behavioural changes in water bodies. These species can also tell about cumulative effects of pollutants in the aquatic ecosystem (Cuadro *et al.,* 2022). In an aquatic ecosystem, the life of biota closely depends upon the physical, chemical, and biological properties of water, all these directly act as a controlling factor. Planktons respond rapidly to ecological changes and act as an excellent indicator of water quality and trophic conditions due to their short time and rapid rate of reproduction. Under natural conditions, the occurrence of planktons is identified with the optimum range in relation to abiotic components (Oxygen fixation, Temperature and pH) as well as the biotic connections among organisms. The changes that occur in communities of planktons help to determine the trophic state of water bodies (Rani *et al.,* 2021).

Climate change, caused by an increase in atmospheric carbon dioxide (CO_2) over the past centuries, has already caused significant shifts in marine and terrestrial ecosystems (Ingels *et al.,* 2014). The direct consequences of increasing CO_2 include the increase in ocean temperature (Bindoff *et al.,* 2007) and acidity (Doney *et al.,* 2012). The ocean temperature is warming, the sea level is rising and sea surface salinity has increased (Durack and Wijffels, 2010). CO_2 concentrations in the atmosphere have increased primarily due to fossil fuel combustion and carbon emissions from land (IPCC, 2014). Climate change probably affects the number of hurricanes and causes changes in current systems, leading to the mass destruction of habitats. The speed of the current climate change is comparable with other dramatic episodes in our planet's history that led to mass extinction events (Van De Schootbrugge and Gollner, 2013). From coastal to deep-sea ecosystems, ecologists are exploring new areas of research to find model organisms that help predict the future of life on our planet. Among the different categories of organisms, meiofauna offer several advantages for the study of marine benthic ecosystems. Meiofauna is a collective name for one of the most diversified communities of the marine realm (encompassing 24 of the 35 animal phyla), including small organisms, unicellular protists and multicellular metazoans that live in aquatic sediments (Giere, 2009). Four taxonomic groups are valuable for predicting global changes: foraminifers (especially calcareous forms), nematodes, copepods and ostracods. Environmental variables are fundamental in the interpretation of meiofaunal patterns and multistressor experiments are more informative than single stressor ones, revealing complex ecological and biological interactions. Global change has a general negative effect on meiofauna, with important consequences on benthic food webs. However, some meiofaunal species can be favoured by the extreme conditions induced by global change, as they can exhibit remarkable physiological adaptations. Due to their high abundance and diversity, their widespread distribution, their rapid generation time and fast metabolic rates, meiofaunal organisms are vital contributors to ecosystem function, including nutrient cycling and the provision of energy to higher trophic levels (Woodward, 2010). These characteristics make meiofaunal organisms excellent candidates to test general ecological hypotheses and theories (Bonaglia *et al.,* 2014).. Meiofauna are characterised by a high sensitivity to environmental changes due to their short generation time and the lack of pelagic larval dispersion for the dominant meiofaunal groups (Boldina *et al.,* 2014). Recent studies have shown that meiofaunal communities respond to warming both in seawater and freshwater ecosystems (Gingold *et al.,* 2013). In freshwaters, macrofaunal and meio-community composition has been shown to change dramatically across a thermal gradient, with indications of low food-web stability in warmer currents (O'Gorman *et al.,* 2012). In subtropical meiofaunal communities, an increase of temperature negatively affects species-rich communities, causing the mortality of dominant species, which may have serious consequences for the benthic food web (Gingold *et al.,* 2013). In the deep eastern Mediterranean Sea, global warming has altered the carbon and nitrogen cycles and has had negative effects on deep-sea bacteria and benthic meiofauna in terms of abundance and diversity (Danovaro, 2010). Furthermore, deep-sea meiofaunal secondary production is expected to decrease in many parts of the ocean due to ocean warming (Mora *et al.,* 2013). Some meiofauna groups, such as foraminifers and ostracods, can be very useful to the study of past climate change by the recovery of their fossil shell. Benthic foraminifera, particularly species with calcareous tests, have a superb fossil record and are widely used by micropalaeontologists to reconstruct conditions in the historical and geological pasts (Schönfeld *et al.,* 2012). The abundance, rapid reproductive rate and ecological sensitivity of species to different environmental conditions also makes foraminifera excellent candidates for monitoring the status of modern ecosystems, particularly those exposed to human impacts (Mojtahid *et al.,* 2008; Schönfeld *et al.,* 2012). Yasuhara *et al.,* (2014) used ostracod and foraminiferal fossil records from the last 250,000 years on the Shatsky Rise in the North Pacific Ocean to demonstrate that rapid climate changes affect deep-sea benthic diversity. Thus, meiofauna reflect changes derived from environmental disturbance, both spatially and temporally (Moreno *et al.,* 2008), and can be considered a collective indicator of environmental quality since they display specific responses to different types of anthropogenic disturbance (Danovaro *et al.,* 2004b). Furthermore, the study of meiofauna is cost-effective compared with that of other benthic components (Rogers *et al.*, 2008). These advantages favour the use of meiofauna as bioindicators, especially when the disturbance source has not been identified. Nevertheless, meiofaunal taxonomic identification remains a challenge and a lack of experts complicates the task.

(ii) Plant Bioindicator

Plants are used as very sensitive tools for prediction and recognition of environmental stresses. The presence or absence of some specific plants or other vegetation provides ample information about environmental health. Among the lower plants taxa involved as bioindicators include a wide range of lichens, algae, fungi (leaf yeasts), etc. which help in the detection of both short term as well as long term effects based on appropriate choice of organisms. For example, lichens are considered ideal bioindicators or monitoring agents due to their vulnerability and resilience to different environmental effects. Lichen, the mutualism between algae and fungi mainly grows on the trunks of forest plants. The slighter modification in forest air quality, climate and structure can be known by the reaction of lichen. Sometimes, the lichens disappear from the forest ecosystem due to environmental stress like increase in toxic gas content (Khatri and Tyagi, 2015).

Apart from being excellent accumulators of pollutants (heavy metals, aromatic hydrocarbons and nitrogen), lichens also indicate land-use change and various other anthropogenic activities world-wide (Zhao *et al.*, 2019; Wang *et al.*, 2020). Lichens are also a proven indicator of climate change (Sancho *et al.*, 2017; Ellis, 2019). Unlike pollution monitoring studies, climate change studies are influenced by other ecological factors, such as phorophyte types, forest structure and pollution loads. Therefore, climate changes studies employing experimental, and modelling systems must incorporate different aspects of the environment (Alatalo *et al.*, 2017; Ellis 2019). In Antarctica these interfering factors are minimized, especially in lichen-based climate changes studies as the continent has minimal air pollution (Mietelski *et al.*, 2000; Bargagli *et al.*, 2004) and anthropogenic influence but has diverse lichen communities and environmental gradients (Ovstedal and Smith, 2001).

Antarctica offers several advantages for climate change studies using lichens as integrative bioindicators due to its unique global ecology positioning. The minimal human interference, lowest pollution levels, dominant ground vegetation of lichens with minimal to no vegetative competition, a wide range of climatic with gradients of temperature and precipitation are some of these advantages. However, studies carried out so far are scarce in comparison to other habitats of the planet. The lichen communities of Antarctica show diverse behaviour in the pattern of distribution and physiology as an indicator to natural climatic gradients and experimental warming, which can be summarised as follows: 1) The lichen communities in maritime Antarctica show high diversity due to milder temperature regime and high precipitation, which tend to extend southwards with climate warming. 2) The cyanolichens are limited to maritime Antarctica and peninsular regions. Climate change-induced warming will increase their extension deep into the continental area with limitations of physiological parameters. 3) Climate change-induced warming will increase the pace of glacier retreat, increasing the current extent of lichen communities in all the habitats of Antarctica. 4) The increased temperature will tend to alter the net photosynthesis of lichens. It will have a detrimental effect on the endemic species as they are less adaptive than lichens having circumpolar distribution. This will lead to the extinction of some species during extension and invasion of other species. The studies carried out so far on Antarctica's lichens indicate that climate change is harmful, leading to biodiversity loss (Nayaka and Rai, 2022).

Climate change affects species composition and dynamics across the planet. The impacts on individual plant species comprise shifts in phenology, productivity, distribution and, thus, biodiversity. Therefore, phenological phases are used as indicators for detecting ecological impacts of climate change on flora and fauna such as plants, migratory birds or fishes and, consequently, on ecosystems reflecting the results of manifold combinations of environmental interactions (Ahas and Asa, 2006). Phenology is the study of periodicity in the life cycle events of plants and influenced by seasonal variations in weather factors (Demaree and Rutishauser, 2012). Changes in the timing of phenological stages of plants such as foliation, flowering, fruit ripening, colour changing, and leaf fall are recognized as globally coherent ecological fingerprints of climate change. Plant development in terms of phenological stages as exemplified is a rather sensitive bio-meteorological response to environmental variation (Holopainen *et al.*, 2013), which is ecologically meaningful since changes in pheno-phases serve as both forcing and inhibiting ecological processes (Parmesan, 2013) across spatial scales from individuals to landscapes (Rosenzweig *et al.*, 2007). Hence, shifts in phenology might be of importance for the implementation of mitigation and adaptation.

Phenological remarks are excellent biological indicators of climate change. Several studies have demonstrated phenological responses to climate change, such as earlier dates for bird migration, changes in species distributions, and earlier dates for flowering. The flowering time of plants is affected by solar radiation, temperature, precipitation, humidity and evaporation, cloudiness, insulation, air pressure, and wind. It has long been recognised that phenological cycles are often strongly influenced by changes in climate, particularly temperature and rainfall because the commencement of each development period for plants requires certain critical and accumulated temperature (Rosenzweig *et al.*,

2008). It is now well documented that plant phenology has an important connection to climatic components. Due to its sensitive response to climate, ecologists are now using plant phenology as an indicator for environmental monitoring, particularly in detecting climate change. It is possible to model the relationship between pheno-phases and climatic elements due to the availability of long term phenological data for several plant and animal species. The study of changes in the timing of plant phenophases in response to climate warming is important for two reasons as:1) It demonstrates to the public and policymakers that climate change is already happening even in response to the current modest increase in temperature. 2) It quantifies the extent of change in natural ecosystems that we can expect in the future. If we want to predict the future response of species to a changing climate, we need to explore how species have responded to the climate in the past (Apoorva and Rai, 2020).

(iii) Animal Bioindicator

Indicator species are often the first in their ecosystem to be affected by a particular environmental change, such as a warming climate, pollution, human development, and other environmental degradation. By monitoring changes in the behaviour, physiology, or number of an indicator species, scientists can monitor the health of its whole environment. Thus, animals offer key clues for environmental changes (Joanna, 2006). To be an effective bioindicator—a species should have some key characteristics. Its health should be a bellwether for the health of other species in the same ecosystem; a species that is uniquely vulnerable or struggling makes a poor bioindicator. And any changes it undergoes should be clear and measurable, acting as an early warning system for broader change. Last, it should respond to change in a predictable way (Rainio and Niemela, 2003).

Amphibians such as frogs and toads are considered indicator species because they're highly sensitive to change in their environments. By monitoring them, scientists can learn about changes in the health of entire ecosystem. Over the last few decades many amphibian populations have plummeted around the world. Dozens of species have vanished entirely; others are simply becoming harder and harder to find. There may be several factors working to harm the world's amphibians. Among these are climate changes, including global warming and thinning of the ozone layer. Destruction of habitat is certainly a problem; when rain forests get chopped down and marshes get filled in, frogs are left with nowhere to go. Pollution is also probably connected to the decline (Beebee and Griffiths 2005). Since frogs absorb water directly through their skin, they're especially vulnerable to water pollutants like pesticides and acid rain. Frogs are considered "indicator species," because they are the first to be affected by degradation of the environment. Their dwindling numbers may be a warning to us that our planet is becoming unliveable. No matter what's behind it, the disappearance of frogs is cause for concern. With a combination of conservation and study, we can help keep frogs hopping for another 190 million years. Micro analytical methods give a new opportunity for using frogs as bioindicators. Frogs have long been considered to be the barometers of the climate as their habitat is small and their populations are at greater threat than species like tigers, which are wider. Studying them provides information about the impact of climate change at micro-level. The long-term monitoring of frog activity therefore will be invaluable in knowing the greater impact of climate change and also saving the species.

As global weather patterns change, some frog species are starting to breed earlier in the year which means that tadpoles are hatching too soon into their environments. Frogs typically breed in the early spring in response to warming local temperatures, emerging from hibernation and migrating to nearby water bodies. However, thanks to warmer winters, frogs are being prompted to do this prematurely. This means that frog embryos will develop too early in the year when the possibility of winter relapses occurring is at its greatest. Winter relapses are potentially lethal to the static frog embryos as, if encased within the frozen water, the embryos are at risk of being mechanically crushed by expanding ice sheets. Additionally, there is growing evidence that early hatching, thanks to earlier breeding, is changing the dynamics of the ecosystems in which amphibians thrive by impeding on interspecies interactions (Blaustein and Johnson, 2003). For example, researchers have found that the Natterjack toad breeding sooner leads to increased competition with tadpoles of the already early breeding common frog whose breeding times have not been altered by climate change. To further increase the complexity of the situation, a warming climate is also known to be responsible for increased precipitation variability. This is most certainly problematic for many frog species due to their heavy reliance on water throughout their life cycles (Simon *et al.,* 2011).

Bats are a species-rich mammal group that provide key services such as pest suppression, pollination of plants of economic importance or seed dispersal. Bats show clear reactions to environmental alterations and as such have been proposed as potentially useful bioindicators. Based on the relatively limited number of studies available, bats are likely excellent indicators in habitats such as rivers, forests, and urban sites. Bats are widespread across the globe and their physiological requirements are greatly influenced by ambient temperature, which affects life cycle characteristics such as hibernation and reproduction (Altringham, 2011). Bats are therefore likely to react to climate change, yet responses have so far been recorded only in a few species [e.g., (Smeraldo *et al.,* 2021)]. The bat's large body surface makes dehydration a considerable risk that must be prevented by drinking nightly, so climate change driven disappearance of drinking sites poses a further threat to bats. This dependence upon climate makes bats excellent candidate organisms to indicate biotic responses to climate change, as proposed by (Newson *et al.*, 2009). This idea is pursued by the CA18107 COST Action "ClimBats" (www.climbats.eu, accessed on 10 May 2021), a European-Union funded action that, among its several goals, aims at designing monitoring networks spread across Europe to monitor compositional changes in bat assemblages such as to reflect responses to climate change. The use of bats as bioindicators has pros and cons. The authors of Mammal Diversity Database 2021 identified eight points on which bases bats would represent excellent bioindicators. These comprise: (a) relative taxonomic stability; (b) wide geographic ranges; (c) rich trophic diversity; (d) provision of key ecosystem services, (e) graded responses to environmental alteration correlated with those of other biodiversity components, such as insects; (f) rapid population declines due to slow population growth; (g) possibility of measuring several variables (population size, feeding activity, etc.); and (h) the role of bats as reservoirs of emerging infectious diseases whose epidemiology could reflect environmental stress. The problems that hinder a standardised use of bats as bioindicators are related to the costs of the technology involved in (especially acoustic) bat surveys and the often-coarse taxonomic resolution provided by echolocation call analysis (Revilla-Martín *et al.*, 2020).

References

Ahas, R., and Aasa, A. 2006. The effects of climate change on the phenology of selected Estonian plant, bird and fish populations. *International Journal of Biometeorology* **51**(1): 17-26.

Alatalo, J. M., Jägerbrand, A. K., Chen, S., and Molau, U. 2017. Responses of lichen communities to 18 years of natural and experimental warming. *Annals of Botany* **120**(1): 159-170.

Allaby, M. A. 2020. Dictionary of Zoology, 5th edition.; Oxford University Press: London, UK, pp: 736.

Altringham, J. D. 2011. Bats: From Evolution to Conservation; Oxford University Press: New York, USA, pp: 331.

Apoorva and Rai, S, K. 2020. Prospects of Phenological Studies in Fruit Crops: A Biological Indicator of Climate Change. *Agriculture and Food:E-Newsletter* **2**(11):384-388.

Appenzeller, C., Begert, M., Zenklusen, E., Scherrer, S, C. 2008. Monitoring climate at Jungfraujoch in the high Swiss Alpine region. Science of the Total Environment 391:262-268.

Bargagli, R., Battisti, E., Focardi, S., & Formichi, P. 2004. Preliminary data on enviornmental distribution of mercury in northern Victoria Land, Antarctica. *Antarctic Science* **5**(1): 3-8.

Beebee, T. C. J. and Griffiths, R. A. 2005. The amphibian decline crisis: A watershed for conservation biology? Biological Conservation, **125**: 271-285.

Bindoff, N. L., Willebrand, J., Artale, V., Cazenave, A., Gregory, J. M., Gulev, S., Hanawa, K., Le Quere, C., Levitus, S., Nojiri, Y. and Shum, C.K., 2007. Observations: oceanic climate change and sea level. pp: 385-428.

Blaustein, A. R., and Johnson, P. T. 2003. The complexity of deformed amphibians. *Frontiers in Ecology and the Environment* 1(2): 87-94.

Boldina, I., Beninger, P. G., and Le Coz, M. 2014. Effect of long-term mechanical perturbation on intertidal soft-bottom meiofaunal community spatial structure. *Journal of Sea Research* **85**: 85-91.

Bonaglia, S., Nascimento, F. J., Bartoli, M., Klawonn, I., and Brüchert, V. 2014. Meiofauna increases bacterial denitrification in marine sediments. *Nature communications* **5**(1): 1-9.

Cash, D. W., Clark, W. C., Alcock, F., Dickson, N. M., Eckley, N., Guston, D. H., Jäger, J. and Mitchell, R. B. 2003. Knowledge systems for sustainable development. *Proceedings of the National Academy of Sciences of the United States of America* **100**: 8086-8091.

Caza, F., Cledon, M., and St-Pierre, Y. 2016. Biomonitoring climate change and pollution in marine ecosystems: a review on Aulacomya ater. *Journal of Marine Biology*.

Cuadro, D. L. C., Berasategui, A. A., and Menéndez, M. C. 2022. Zooplankton: The Ocean Drifters. In Marine Biology a Functional Approach to the Oceans and their Organisms. CRC Press pp: 113-149.

Danovaro, D., Gambi, C., Dell'Anno, A., Corinaldesi, C., Fraschetti,S. 2008. Exponential decline of deep-sea ecosystem functioning linked to benthic biodiversity loss. *Current Biology* **18**:1-8.

Demarée, G. R., Rutishauser, T. 2011. From "Periodical Observations" to "Anthochronology" and "Phenology"–the scientific debate between Adolphe Quetelet and Charles Morren on the origin of the word "Phenology". *International journal of biometeorology* 55(6), 753-761.

Doney, S, C., Ruckelshaus, M., Duffy, J, E., Barry, J, P., Chan, F. 2012. Climate change impacts on marine ecosystems. *Annual Review* of Marine Science **4**:11–37.

Durack, P. J., and Wijffels, S. E. 2010. Fifty-year trends in global ocean salinities and their relationship to broad-scale warming. *Journal of Climate* **23**(16): 4342-4362.

Ellis, C. J. 2019. Climate change, bioclimatic models and the risk to lichen diversity. *Diversity* **11**(4): 54.

Giere, O. 2008. Meiobenthology: the microscopic motile fauna of aquatic sediments. Springer Science and Business Media.

Gingold, R., Moens, T., and Rocha-Olivares, A. 2013. Assessing the response of nematode communities to climate change-driven warming: a microcosm experiment. *PLoS One* **8**(6): e66653.

Guralnick, J. 2002. Biological Indicators as Early Warning of ESNO Events. *Regional Disaster Information Centre (CRID).*

Holopainen, J., Helama, S., Lappalainen, H., and Gregow, H. 2013. Plant phenological records in northern Finland since the 18th century as retrieved from databases, archives and diaries for biometeorological research. *International Journal of Biometeorology* **57**(3): 423-435.

Ingels, J., Dashfield, S. L., Somerfield, P. J., Widdicombe, S., and Austen, M. C. 2014. Interactions between multiple large macrofauna species and nematode communities—Mechanisms for indirect impacts of trawling disturbance. *Journal of Experimental Marine Biology and Ecology* 456: 41-49.

IPCC. 2014. Climate change 2014: impacts, adaptation, and vulnerability. Cambridge University Press, Cambridge.

Joanna, B. 2006. Bioindicators: types, development, and use in ecological assessment and research. *Environmental Bioindicators* **1**(1): 22-39.

Khatri, N., and Tyagi, S. 2015. Influences of natural and anthropogenic factors on surface and groundwater quality in rural and urban areas. *Frontiers in life science* **8**(1): 23-39.

Mammal Diversity Database. Mammal Diversity Database (Version 1.4), [Data download May/2021]

Manning, W. J. 1998. The use of plants as bioindicators of ozone. on Air Pollution and Climate Change Effects on Forest Ecosystems 19.

Middleton, J. T., Kendrick, J. B., and Schwalm, H. W. 1950. Injury to herbaceous plants by smog or air pollution. *Plant Disease Reporter* **34**(9).

Mietelski, J., Gaca, P., and Olech, M. 2000. Radioactive contamination of lichens and mosses collected in South Shetlands and Antarctic Peninsula. *Journal of Radioanalytical and Nuclear Chemistry* **245**(3): 527-537.

Mojtahid, M., Jorissen, F., and Pearson, T. H. 2008. Comparison of benthic foraminiferal and macrofaunal responses to organic pollution in the Firth of Clyde (Scotland). *Marine Pollution Bulletin* **56**(1): 42-76.

Mora, C., Wei, C. L., Rollo, A., Amaro, T., Baco, A. R., Billett, D., Bopp, L., Chen, Q., Collier, M., Danovaro, R. and Gooday, A. J., 2013. Biotic and human vulnerability to projected changes in ocean biogeochemistry over the 21st century. *PLoS biology* **11**(10): e1001682.

Moreno, C. E., Sánchez-Rojas, G., Pineda, E., and Escobar, F. 2007. Shortcuts for biodiversity evaluation: a review of terminology and recommendations for the use of target groups, bioindicators and surrogates. *International Journal of Environment and Health* **1**(1): 71-86.

Moreno, M., Vezzulli, L., Marin, V., Laconi, P., Albertelli, G., and Fabiano, M. 2008. The use of meiofauna diversity as an indicator of pollution in harbours. *ICES Journal of Marine Science* **65**(8): 1428-1435.

Nayaka, S., and Rai, H. 2022. Antarctic Lichen Response to Climate Change: Evidence from Natural Gradients and Temperature Enchantment Experiments. In Assessing the Antarctic Environment from a Climate Change Perspective pp: 235-253.

Newson, S. E., Mendes, S., Crick, H. Q., Dulvy, N. K., Houghton, J. D., Hays, G. C.; Robinson, R. A. 2009. Indicators of the impact of climate change on migratory species. Endangered Species *Research* **7**: 101-113. [CrossRef]

Niemelä, J. 2000. Biodiversity monitoring for decision-making. *In Annales Zoologici Fennici* pp: 307-317.

Noss, R. F. 1990. Indicators for monitoring biodiversity: a hierarchical approach. *Conservation biology* **4**(4): 355-364.

O'Gorman, E. J., Pichler, D. E., Adams, G., Benstead, J. P., Cohen, H., Craig, N., Cross, W. F., Demars, B.O., Friberg, N., Gislason, G. M. and Gudmundsdottir, R. 2012. Impacts of warming on the structure and functioning of aquatic communities: individual-to ecosystem-level responses. In *Advances in ecological research* **47**:81-176.

Øvstedal, D. O. and Smith, R. L., 2001. Lichens of Antarctica and South Georgia: a guide to their identification and ecology. Cambridge University Press.

Parmesan, C. 2007. Influences of species, latitudes and methodologies on estimates of phenological response to global warming. *Global change biology* **13**(9): 1860-1872.

Pereira, H. M., Junker, J., Fernández, N., Maes, J., Beja, P., Bonn, A., Breeze, T., Brotons, L., Bruelheide, H., Buchhorn, M. and Capinha, C. 2022. Europa Biodiversity Observation Network: integrating data streams to support policy. *ARPHA Preprints* **2**: .e81207.

Rainio, J., and Niemelä, J. 2003. Ground beetles (Coleoptera: Carabidae) as bioindicators. *Biodiversity & Conservation* **12**(3): 487-506.

Rani, S., Gunjyal, N., Ojha, C. S. P., and Singh, R. P. 2021. Review of challenges for algae-based wastewater treatment: Strain selection, wastewater characteristics, abiotic, and biotic factors. *Journal of Hazardous, Toxic, and Radioactive Waste* **25**(2): 03120004.

Revilla-Martín, N., Budinski, I., Puig-Montserrat, X., Flaquer, C. and López-Baucells, A. 2020. Monitoring cave-dwelling bats using remote passive acoustic detectors: a new approach for cave monitoring. *Bioacoustics* **30**(5): 527-542.

Rockstrom, J., Steffen, w., Noone, K., Persson, A. Chapin m, FS, Lambin, E., Lenton, TM, Scheffer, M., Folke, *C., and Schellnhuber,* H. 2009. Planetary boundaries: Exploring the safe operating space for hurnaoity, *Ecol. Soc* 14: 32.

Rogers, S. I., Somerfield, P. J., Schratzberger, M., Warwick, R., Maxwell, T. A., and Ellis, J. R. 2008. Sampling strategies to evaluate the status of offshore soft sediment assemblages. *Marine Pollution Bulletin* 56(5), 880-894.

Rosenzweig, C., Casassa, G., Karoly, D. J., Imeson, A., Liu, C., Menzel, A., Rawlins, S., Root, T. L., Seguin, B., Tryjanowski, P. and Parry, M. L., 2007. Assessment of observed changes and responses in natural and managed systems.

Rosenzweig, C., Karoly, D., Vicarelli, M., Neofotis, P., Wu, Q., Casassa, G., Menzel, A., Root, T. L., Estrella, N., Seguin, B. and Tryjanowski, P., 2008. Attributing physical and biological impacts to anthropogenic climate change. *Nature* **453**(7193): 353-357.

Sancho, L. G., Pintado, A., Navarro, F., Ramos, M., De Pablo, M. A., Blanquer, J. M., Raggio, J., Valladares, F. and Green, T. G. A., 2017. Recent warming and cooling in the Antarctic Peninsula region has rapid and large effects on lichen vegetation. *Scientific Reports* **7**(1): 1-8.

Schönfeld, J., Alve, E., Geslin, E., Jorissen, F., Korsun, S., and Spezzaferri, S. 2012. The FOBIMO (FOraminiferal BIo-MOnitoring) initiative—Towards a standardised protocol for soft-bottom benthic foraminiferal monitoring studies. *Marine Micropaleontology* **94**, 1-13.

Simon, E., Puky, M., Braun, M., and Tóthmérész, B. 2011. Frogs and toads as biological indicators in environmental assessment. *Frogs: Biology, ecology and uses* pp: 141-150.

Smeraldo, S., Bosso, L., Salinas-Ramos, V. B., Ancillotto, L., Sánchez-Cordero, V., Gazaryan, S., and Russo, D. 2021. Generalists yet different: Distributional responses to climate change may vary in opportunistic bat species sharing similar ecological traits. *Mammal Review* **51**(4): 571-584.

Van De Schootbrugge, B., and Gollner, S. 2013. Altered primary production during mass-extinction events. *The Paleontological Society Papers* **19**, 87-114.

Wang, C. H., Hou, R., Wang, M., He, G., Li, B. G., and Pan, R. L. 2020. Effects of wet atmospheric nitrogen deposition on epiphytic lichens in the subtropical forests of Central China: Evaluation of the lichen food supply and quality of two endangered primates. *Ecotoxicology and Environmental Safety* **190**: 110128

Watson, R., Chung, L., Gitay, H., Herold, A., Kelleher, S., Kumari, K., Lamb, R., Lantheaume, F., Ploetz, C., Sivakumar, M. V. K. and Watt, A. 2003. Approaches for supporting planning, decision making and public discussions. In Interlinkages between Biological Diversity and Climate Change; Advice on the Integration of Biodiversity Considerations into the Implementation of the United Nations Framework Convention on Climate Change and Its Kyoto Protocol. *CBD Technical Series* 10 pp. 88-110).

Woodward, G. 2010. Integrative ecology: from molecules to ecosystems. Academic Press.

Yasuhara, M., Okahashi, H., Cronin, T. M., Rasmussen, T. L., and Hunt, G. 2014. Response of deep-sea biodiversity to abrupt deglacial and Holocene climate changes in the North Atlantic Ocean. *Global Ecology and Biogeograph* **23**(9): 957-967.

Zhao, L., Zhang, C., Jia, S., Liu, Q., Chen, Q., Li, X., Liu, X., Wu, Q., Zhao, L. and Liu, H. 2019. Element bioaccumulation in lichens transplanted along two roads: The source and integration time of elements. *Ecological Indicators* **99** pp:101-107.

Chapter 7

Working of Joint Forest Management (JFM) Committees in Jammu & Kashmir

Shah Murtaza Mushtaq*, Akhlaq Amin Wani, AmirBhat, Talib Bashir, and M A Islam

Division of Natural Resource Management, Faculty of Forestry SKUAST- Kashmir

Email:smurtaza@rediffmail.com

Abstract

National Forest Policy 1988, recommended that 33% of the Geographical area should be under forest cover to achieve ecological balance and harmony. However, only 20.43% of India's geographical area has been assessed as forest area (Forest Survey of India, 1991). Based at the present level of consumption of forest product and on current productivity of forests, has estimated that the countries need a minimum of 0.47 ha. Of forestland to meet the basic needs of the people. JFM is a major forestry programme of the country and therefore Government of India took a number of initiatives to make it sustainable in a long run. Notable among such steps were creations of JFM monitoring cell in the Ministry. In yet another initiative government approved the scheme of Forest Development Agencies (FDA's) which are federation of JFM committees (Bhattacharya *et.al*, 2010). Joint Forest Management programme envisaging different models for sharing of benefits through collection of NTFP's and for timber and bamboo harvesting. Over 84,000 JFMC's covering an area of about 18 million hectares were reported (Bahugana and Upadhya, 2004).

Introduction

The foresters had long been working with the forest dwellers through the taungya system of cultivation under which the agriculture crops were raised alongwith the forest plantations for a few years. Joint Forest Management (JFM) programme in the present form can be traced to the Arabari experiment initiated by foresters in the state of West Bengal. This experiment provided a strong feedback for incorporation of the system in the National Forest Policy of 1988. In many locations people's voluntary groups were engaged in protection of forests without any initiative from the Government. Subsequently, based on the experience, the process of institutionalizing people's participation in forest protection and regeneration began. This type of collective endeavour in protection and management of forests through people's involvement was later termed as Joint Forest Management.

The State of Jammu & Kashmir lies to the extreme North of the country situated between 32°17' and 36°58' North Latitude and 72°26' and 83°30' East Longitude, comprising of an area of 2,22,236 km². The elevation above sea level is 1660 meters and climate of the area falls under temperate climatic regimes with mean annual rainfall of 660 mm and mean temperature of 13.3°C. Maximum temperature of the area may touch to 35°C in summer and the minimum temperature may drop to -10°C in winter months. In the area generally alluvial soils dominate which generally suffer from impeded drainage.

The **Joint Forest Management (JFM)** was formally adopted in Jammu & Kashmir State by issuing a notification SRO 61 dated 19-03-1992. But the State of Jammu & Kashmir had already introduced the concept way back in 1982 in its World Bank aided Social Forestry Project. The project envisaged sharing of produce with the local communities. But the idea did not catch up then and JFM was not fully understood. The practice ended up by signing agreements with panchayat but the exact procedure for involvement of people and their role could not be clearly defined. The system totally collapsed in the year 1984 when the panchayat in the State were dissolved, which had till then served as a link between the people and the department. This was, therefore, replaced by Village Forest Committees (VFC's) in the year 1987-88.Consequent to the guidelines issued by Government of India in the year 1990 and the issuance of SRO 61 of 19-03-1992 and SRO 17 of 12-01-1990 by the State Government, the concept of JFM caught up in the State through formation of VFC's. At present 1599 VFC's are functioning over the State, both with the social forestry project and the territorial wings of the Forest Department and around 10000 hectares of land is under rehabilitation with the assistance and participation of VFC's.

Ongoing Schemes

A. Plan Developmental Schemes

The schematic breakup of proposed requisition of funds is as follows:-

Rehabilitation of Degraded Forests

This is the major Plan Scheme and aims at rehabilitation of degraded forests in the State. It is estimated that more than 10 lakh hectares of Forest area is degraded in the State. Through this scheme efforts are afoot to rehabilitate the degraded and denuded Forest land on scientific lines by way of plantation, construction of check-dams, water harvesting structures and regulating natural regeneration. An amount of Rs. 182.00 lakh has been approved for 2009-10 under state sector and an amount of Rs. 270.25 lakh been approved under District Sector for rehabilitation of Degraded Forests. For the next year 2010-11 an amount of Rs. 208.00 lakh is being proposed for the said scheme @ Rs. 8.00 lakh per Division under State Sector. And Rs. 297.28 lakh has been proposed under District Sector.

Research Education and Training

This scheme aims at providing requisite training to the field staff of Forest Department and provides for filling up of infrastructural gaps in respect of three training schools viz KFT School Chitternar, SC Training School, Miransahib and Forest guards Training School Doomi. A provision of Rs. 7.99 lakh has been kept for 2009-10 for conducting of workshops, extensive Training programmes etc besides supplementing for infrastructural facilities. For the next year 2010-11 an amount of Rs. 8.50 lakh has been proposed for the said scheme

Working Plan and Research

Activities of five divisions, including three Working Plan Divisions, one Photo interpretation Division and One Resources Survey Division, are covered under this scheme. The main activities of all these divisions are concerned for preparation of Working Plans of the forest areas which indicative of overall Forest health of the state. The working plans are useful for researcher's policy makers and academicians etc. The GIS Laboratory established two years back has become functional now. There is therefore a requirement of equipping it with necessary hardware and software for which liberal funding is the requirement. However keeping the approved ceiling in view only an amount of Rs. 8.00 lakh could be earmarked under this scheme for 2009-10, for the next year 2010-11 an amount of Rs. 8.00 lakh has been proposed for the said scheme.

Consolidation and Demarcation

Forest area is being lost due to lack of proper demarcation of its boundaries on ground. To ensure protection of existing forest areas, the department through three demarcation divisions, one in Kashmir and two in Jammu conducts surveys and demarcates the existing forests by putting up the permanent boundary pillars along the fringe lines and prepare records pertaining to forest land by mutation in the revenue records. An amount of Rs. 15.00 lakh has been kept for this activity for 2009-10. During 2010-11 an amount of Rs. 20.00 lakh has been proposed for the implementation of the scheme.

Development of Minor Forest Produce including Medicinal Plants (M.F.P)

It has been felt that timber production has received much attention and development of other Minor Forest produce has virtually been neglected so far. Mistreatment of Minor Forest Produce continued at an alarming rate and the conservation efforts were not apace with it with the result it received a setback. The forest resources in J&K have a tremendous potential of Minor Forest Produce for becoming an independent industry provided adequate funding is made available. The cultivation of medicinal plants and its conservation is another area that needs focused attention. In order to give necessary impetus to the development of Minor Forest Produce including medicinal plants as a full-fledged industry and to see that marketability of the produce is ensured, a provision of Rs. 30.00 lakh has been approved for 2009-10 for the purpose. An amount of Rs. 32.00 lakh has been proposed for the next financial year 2010-11.

Eco Task Force

The Eco Task Force Project (E.T.F) is a joint venture of J&K Forest Department and Territorial Army. Initially, it was launched in the year 1988 to rehabilitate the Rue Catchments in Samba Town of Jammu province. The Territorial Army provides the man power by hiring the services of ex-Army personnel and the funds for the purpose to them are provided by Ministry of Environment & Forests. The Forest Department provides the expert guidance's and the planting materials. Due to inadequate financial resources the project received a setback though it stands awarded 1994 Indira Priyadarshani Vriksha Mitra Award by Government of India. The project was approved for implementation up to 10th plan period viz; 03/2007. However on the insistence of the Army and the locals the extension of the project has been approved till end of 11th Five Year Plan. As per the approved Work Plan 300 Ha of area is required to be treated every year, the estimated cost for afforestation is 1500 Ha, Rs. 465.00 lakh (Every year afforestation of 300 Ha). In view of the revised wage rates cost of implementation at present shall be much more than what has been approved earlier. Since a part of the working season is already over it is estimated that the department will be in a position to incur Rs. 99.00 lakh only. It is because of this reason an amount of Rs. 99.00 lakh has been anticipated to be incurred during 2009-10. The revision is proposed as per the directives of Administrative Department conveyed vide No: FST/Plan-36/95 dated 04.01.2008 where-under extension of the project against the Action Plan has been approved. Rs. 110.00 lakh has been proposed for the next year 2010-11.

CM's Participatory Afforestation Scheme

In order to involve people in the regeneration of degraded forests and increasing the tree coverage on vacant lands, the scheme ***"ChiefMinisters Participatory Afforestation Programme"*** has been launched from the year 2004-05 on pilot basis. The objective of the scheme is to make tree plantation on open forest lands, vacant lands and other Government lands by giving its full benefits to the beneficiaries involved in the programme. The Forest Department provide the infrastructural support viz. planting materials, fencing materials and expert guidance to the beneficiaries. The scheme is being implemented in 40 forest divisions including 12 divisions of Social Forestry Department. An amount of Rs. 4.00 lakh per division is provided to each division for implementation of the scheme. Accordingly an allocation of Rs. 160.00 lakh has been approved for 2009-10. An amount of Rs. 180.00 lakh has been proposed for the next financial year 2010-11 @ Rs. 4.50 lakh for 40 Division including 12 Division of Social Forestry Department.

11th Five Year Plan under NAP (2007- 2012)

- **1107 JFMCs have been** formed in **27 Forest Development Agency**
- A total of **53157 ha was** targeted to be achieved from (2007-12) with a financial support of **Rs. 114.88 crore**
- Of this only an amount of **Rs. 1916.17 lakh** was released during 2007-8 and 2008-09 against which an area of **14161 Ha** has been registered to have been achieved with an expenditure of **Rs. 14.10 crore** during the corresponding period.

Methodology

An integrated approach was adapted to study the Methodology of project planning and implementation by reviewing available literature and by interviewing the people involved with the project at various stages. The villages were randomly selected, Socio-economic survey, semi structured interviews and questionnaires.

Collection of Data

1. Collection of Primary Data

Primary data was collected with the help of well – structured questionnaires, interviews and observations. The questionnaires was designed to meet the objectives of the study, tested in the field and standardized for the purpose. The questionnaire had open-ended questions regarding awareness about JFM and forest resource utilization. Participatory Rural Appraisal (PRA) and Rapid Rural Appraisal (RRA) were used as a tool for the data collection. Following PRA/RRA Methods were used

a. Transect walk
b. Participatory mapping
c. Livelihood analysis
d. Interviews

2. Collection of Secondary Data

The secondary data was collected from the various records and project reports of the Forest Department. To get information on demography, administrative set up and ecological data, several records of J&K Forest Department were consulted. Information on local demands, Entry point activities, Project area, Project planning, Project funding, Implementing agencies, Structure – Composition and activities of Forest Development Agencies were collected from the various records & project reports of Forest Department.

Result and discussion

Anantnag Forest Division

The Anantnag Forest Division covers an forest area of 1034 sq. Km's, spread over a vast tract and is situated between 33° - 22' and 34° - 15' North Latitude and 74° - 48' and 75° - 30' East Longitude. Occupying the south -eastern portion of Kashmir Valley, the tract lies in Anantnag District of Kashmir Province. It is bounded on the east and the south by the lofty Pir Panjal mountain range, separating it from the Ramban and Kishtwar Forest Divisions. The Anantnag Forest Division comprises of five Ranges having 28 Village Forest Committees. Large scale deforestation over the past 2-3 decades during the most turbulent phase of the disturbance has rendered many beats and compartments in the division denuded and degraded. In order to rehabilitate these degraded forests this scheme had been started in 2001.

Achievements of FDA Anantnag from 2001-2009

- Natural Regeneration

  Under this scheme 317.75 Hectare was taken up for fencing and plantation in different villages.

- Management Intervention & Raising of Nursery

  Two nurseries were raised under this scheme viz Kokernag and Momandanji where quality plant material indigenous and exotic species were raised.

  An amount of Rs 12.95 Lakhs was involved for raising of Nurseries at Kokernag at 8.75 and at Momandanji at a cost of Rs 4.19 Lakhs.

- Silvi pasture Development

  Under this scheme 77 hectare area were tackled where plantation and fencing work was taken at a cost of 5.306 Lakhs.

- Soil & Moisture Conservation

  Under this scheme construction of Check Dam work was taken up to protect soil erosion. 1940 Cu M of bunds were taken up in different VFC's.

- Entry Point Activity

  Under this scheme minor engineering works like construction of bathroom, culverts, paths, tube wells, were constructed.

FDA Lidder Forest Division, Bijbehara

Lidder Forest Division comprise of five ranges i.e., Pahalgam, Mattan, Bijbehara, Tral, Pampore. Due to the abnormal conditions in the valley the forests were badly affected causing huge loss to the forests. In order to rehabilitate these degraded forests under this scheme the project period has given its start from 2007-08 for five years (Upto 2011-12).

Table 1: Statement showing Physical and Financial targets during 2007-08 to 2011-12 under National Afforestation Programme through FDA Lidder Forest Division

S No	Components	Physical Targets	Financial Targets(Lakhs
01	Afforestation works	1104 Hectare	Rs 213.521
02	Infra Structure	11 No's	Rs 22.00
03	Maintenance/BUC	1104 Hectare	Rs26.40
04	Watch and Wards	--	Rs13.86
	Total		Rs275.78

Table 2: Statement showing year wise area (in hectare) treated with effect from 2007-08 to 2011-12

Period	Target Area (in hectare)	Achieved Area (in hectare)
2007-08	276	244
2008-09	242	240
2009-10	212	-
2010-11	196	-
2011-12	178	-
Total	1104	

Table 3: Physical/ Financial targets with effect from 2007-08 to 2011-12 and maintenance upto the year 2013-14

Year	Targets	Afforestation Works	Infra structure	Maintenance	Watch and Ward	Total
2007-08	Physical (in ha)	276	11 No's	--	--	276
	Financial (in lakhs)	44.23	22.00	--	2.772	69.002
2008-09	Physical (in ha)	242	--	--	--	242
	Financial (in lakhs)	51.11	--	6.60	2.772	60.482
2009-10	Physical (in ha)	212	--	--	--	212
	Financial (in lakhs)	43.58	--	6.60	2.772	52.942
2010-11	Physical (in ha)	196	--	--	--	196
	Financial (in lakhs)	39.57	--	6.60	2.772	48.942
2011-12	Physical (in ha)	178	--	--	--	178
	Financial (in lakhs)	35.03	--	6.60	2.772	42.402
Total	Physical (in ha)	1104	11 No's	--	--	1104
	Financial (in lakhs)	213.52	22.00	26.40	13.86	275.78

FDA Marwah Forest Division

Table 4: Marwah Forest Division Consists of 3 ranges Viz Dachhan, Marwah and Udil.

S.No	Range	No of VFC's
1	Dachhan	10
2	Marwah	05
3	Udil	19
	Total	34

Achievements

- Natural Regeneration

  Under this scheme an area of 427 hectare was under taken up for fencing and plantation in different villages by the respective VFC's.

- Management Intervention

  Under this scheme an area of 196 hectare was taken up for fencing and plantation at an amount of Rs 19.98 lakhs. 195,050 plants were planted in different compartments.

- Silvi Pasture Development

  Under this Scheme 177 hectare area were tackled where plantation and fencing work was taken up at a cost of 15.306 Lakhs.

- Entry Point Activity

  Under this scheme 12 training camps were organized at different villages.

In addition to this minor engineering works like culvert at Kharian, Path from Passerkote to Takpura (1m × 1.5Km), Toilet/ Bathroom at Wanipora Passerkote were constructed. Total expenditure of 8.00 Lakhs was incurred in this scheme.

FDA Kamraj Forest Division

The Kamraj Forest Division comprise of five Ranges viz North-lolab, South-lolab, Kandi, Naihari and Kupwara having 58 Village Forest Committees.

Table 5: Scheme Wise Details of Afforestation Works F.D.A, Kamraj Forest Division

S. No	Item	Natural Regeneration	Artificial Regeneration	Pasture Development	Total Area (in ha)
1	No. of V.F.C's.	58	58	58	—
2	Total Area (in ha)	270	960	270	1500
3	Area Per V.F.C (in ha)	4.65	16.55	4.65	25.85
4	Area covered during 2004-05 (in ha)	118	304	85	507
5	Area covered during 2005-06 (in ha)	65	246	70	381
6	Area covered during 2006-07 (in ha)	43	218	60	321
7	Area covered during 2007-08(in ha)	44	192	55	291

Conclusion

- Reorientation and training of the Forest Department staff in JFM approach is the most crucial input needed to make this programme a success. Frequent transfers of the staff result in wastage of training efforts and have to be avoided.
- Mechanisms need to be evolved to make JFM a self-sustaining activity within a short a period as possible by ensuring financial viability of VFC's.
- JFM training should be started in the Forest Training Schools so that the new entrants are exposed to this new approach before entering the service.
- At least 50% members of the JFM general body should be women.
- At least 33% of the membership in the JFM Executive Committee should be filled from amongst the women member. A women member of the committee should fill one of the office bearer i.e. president/vice-president/ secretaries.

References

Annual Administrative Report (2009). Forest Statistics Division, J&K Forest Department.

Bahugana, V. K. and A. Upadhyay (2004). Monitoring needs for JFM: The perspective of policy makers. In: Bahugana, V. K., Mitra, K., Capistrano.D and Saigal, S. (eds.). *Root to Canopy: Regenerating Forests through Community-State partnerships*. Winrock International India/ Commonwealth Forestry Association-India Chapter, New Delhi. Pp: 309-316.

Baig, M. A. (2008). *Project Report for Forest Development Agency*. Anantnag Forest Division, South Kashmir.

Bhattacharya, P., K. N. Krishna Kumar, R. Prasad, K. C. Malhotra, D. Debnath, G. Yadav, B. K. Prasad, S. Roy, P. Lolita and A. Singh, 2010. *JFM at a Crossroads: Future Strategy and Action Programme for Institutionalizing Community Forestry*. International Centre for Community Forestry, Indian Institute of Forest Management (IIFM), Bhopal. Pp: 1-2.

Digest of Forest Statistics (2009). Forest Statistics Division, Srinagar, J&K Forest Department.

Fotidar, A.N. (1989). Forestry in Jammu and Kashmir: A brief critical review. *Indian Forestry,* **115(6):** pp392-396.

Gill, A.S. & M. M., Roy (2001). Improved silvopastures for Joint Forest Management in India. *Indian Forester*, **127 (3)**: pp 303-314.

Hill, I. and D. Shields (2009). *Incentives for joint forest management in India: analytical methods and case studies*. World Bank Technical Paper No 394. World Bank, Washington, DC., USA. Pp: 111.

Jattan, S.S. and Pratima. (1999). Participatory Forest Management to meet basic needs of forest dependent committees. *Indian Forester*, **125 (9)**: pp 895-903.

Jattan, S.S. and Pratima. (2001). Participatory Forest Management objectives in India. *Indian Forester*, **127(5):** pp 505-511.

Kawosa, M.A. (2001). Towards a pragmatic forest policy, *Forests of Kashmir***:** pp 90-102.

NAEB (2006). *Operational Guideline for the Tenth Five Year Plan*. National Afforestation and Eco-development Board. Ministry of Environment and Forests, New Delhi.

NAEB (2009). *Revised Operational Guideline for the Tenth Five Year Plan*. National Afforestation and Eco-development Board. Ministry of Environment and Forests, New Delhi.

Negi, S.S. (2000). Micro plans, *Joint Forest Management*. International Book Distributors, Dehradun, India. pp: 133-137.

Chapter 8

Ziziphus jujuba in Kashmir - A super fruit with medicinal properties

M. Maqbool Rather, Firdous Shiekh, Ashfaq A Mir and Megna Rashid*

**Faculty of Forestry, SKUAST-Kashmir*

**maqboolfri@gmail.com*

Abstract

Jujube (*Ziziphus jujuba* Mill.), or Chinese date, is the most important species of Rhamnaceae family and is one of the oldest cultivated fruit trees in the world. Chinese jujube is new fruit specie with a high potential to be planted on arid and semiarid areas on marginal, poor and even salty soils. Jujube tree is drought and frost tolerant with wide adaptation. Considering the complex composition of sugars, vitamins, amino acids, etc. jujube fruits have an important nutraceutical effect, being extremely useful for the human healthy diet. The plant can be successfully propagated both by seeds and by vegetative ways. Even the grafting can be applied with good results, for the fast propagation of the new varieties, *in vitro* multiplication is recommended.

Key words: *Ziziphus jujuba*, medicinal properties, drought resistant, propagation, Kashmir Karewas.

Introduction

Plants are a rich source of phytochemicals that occur naturally in different parts like fruits, seeds, flowers, leaves, stems, barks and roots (Liu and Wang, 2019). Nowadays, plant based products, containing a broad spectrum of compounds, are of great interest not only to scientists but also consumers (Guo and Shan, 2010) due to their natural origin and kind of availability (Hu, 2011). It was estimated that the global market of plant extracts accounted for USD 30.8 billion in 2021 and is expected to reach USD 55.3 billion by 2026 (Liu and Wang, 2009). Medicinal and aromatic plants as components of cosmetics, food, and medicinal and health products (Liu *et al.,* 2016) are widely used for aromatic, culinary and therapeutic purposes. Moreover, plant-based extracts are incorporated into natural products that are used in crop cultivation in order to increase their quality and tolerance to environmental stresses (Stanica, 2019). Primary plant constituents e.g., sugars, proteins, amino acids, chlorophyll, and purines and pyrimidines of nucleic acids are principally nutritional substances of plants and, in most cases, do not possess medicinal properties. Secondary constituents (secondary metabolites) such as alkaloids, carotenoids, flavonoids, glycosides, lignins, phenolic compounds, terpenoids, saponins, steroids, and tannins are accountable for multifarious biological, ecological, pharmacological, and toxicological activities (Liu *et al.*, 2016). Most of these secondary metabolites have a role in the defense against environmental factors (e.g., pollution, drought, UV light, pathogens).

Jujube (*Ziziphus jujuba* Mill.) is one of the most important genera of Rhamnaceae family. Its common name is Chinese date or Chinese jujube. In older European texts, it is known as 'common jujube'. It is native to China and distributed in over 47 countries around the world (Li *et al.*, 2005; Liu *et al.*, 2014). The plants are thorny or spinescent shrubs and trees with distinctive tri-nerved leaves, cymose inflorescence, and small perigynous flowers with the ovary surrounded by a thick nectary disc (Islam and Simmons, 2006). All the jujubes have been relatively minor crops, although demand for their production remains steady in many parts of their original domestication (Azam-Ali *et al.*, 2006). It is a medium-sized perennial fruit tree with medicinal properties (commonly diploid species; $2n = 2x = 24$), growing 7–10 meters in height. The tree is very graceful and ornamental in appearance. It has shiny deciduous foliage and produces a fruit with a thin, dark red skin and sweet, white flesh surrounding a stone. Size varies depending on the cultivar and can range from 30 to 120 grams.

Jujuba fruit-A super fruit for future

The fruit is rich in nutrients, and famous for the sweet, apple-like taste with potassium, phosphorus, calcium and manganese being the major mineral components, as well as iron, sodium, zinc and copper (Li *et al.*, 2005). The jujube is a rich source of vitamin C and vitamin B-complex. The vitamin C content is higher than fruits such as oranges. Studies have shown that the anti-oxidant content of fresh jujubes is higher than strawberries, plums, apples, blueberries, blackberries and raspberries (Johnstone, 2012). Jujube fruits are an excellent source of ascorbic acid and carotenoids (Abbas, 1997). The commercial importance of jujube has increased in recent years because of the recognition of the nutritional value of the fruit. The fruits can be consumed fresh or processed into beverage, food, and oriental medicines. Jujube flowers can produce high-quality honey, which made up one third of the Chinese honey production in 2007 and highly valuable source of food and fuel (Azam-Ali *et al.*, 2006). It has been commonly used as a drug in traditional Chinese medicine as an analeptic, palliative, antibechic apart from being used as food, food additive and flavorant for thousands of years. Different parts of jujube such as root, bark, leaf, fruit, and seed have been commonly used as sources of crude drugs from ancient time (Li *et al.*, 2005). In addition, *Z. jujuba* has multiple human health benefits, such as anticancer, antioxidant, anti-inflammatory, anti-obesity, and gastrointestinal protection attributes because of the presence of bioactive compounds in the fruits (Tahergorabi *et al.*, 2015).

The unique growth and fruiting habits of jujube

The biological characteristics of jujube have been studied comprehensively, and a complete theoretical system had been formed by the 1990s. Jujube is highly tolerant to drought and well adapted to diverse climatic conditions and not only survives but produces reasonable yield and growth under severe drought helped by its deep and substantial taproot (Xinguang *et al.*, 2017). It has unique branch and bud characteristics, i.e., usually only the primary shoots can extend, the dormant buds have a very long life, the secondary shoots die back naturally each, the mother-bearing shoots can only extend by ~1 mm per year, and the bearing shoots fall off in the fall. Its flower bud differentiation and fruit set habit are also very distinct, with a short flower bud differentiation time (10 days), a 2-month flowering season, and a low fruit set of only ~1%. Jujube is a frost resistant species and can survive in minimum temperature of −10°C to −20°C with maximum temperature of 36°C and an annual rainfall of 20 cm to 45 cm (Azam-Ali *et al.*, 2006). It is becoming increasingly important because of its wide adaptation, easy management, early bearing, rich nutrition and multiple uses with high degree of tolerance to barren soil thereby proving economical and productive. The jujube has lower water requirements and higher salt tolerance than most of the fruits.

Presence in Valley

The taxon has been found commonly growing on the terraced tablelands, locally called as 'Karewas', and along the foot hills in the Kashmir Valley (Akhter *et al.*, 2013). These lands are typically lacking permanent source of water with dry conditions. This particular species has been found thriving very well under such drastic edaphic as well as climatic conditions with less soil development and less water availability.

Ziziphus jujuba ssp. Spinosa: **Shrubs**; drupes globose to ellipsoid, small (0.6–2 cm across.), orange-red, not shiny, epicarp thick, mesocarp thin and sour, forming a big cavum when dry; stone small, globose to ellipsoid, obtuse at both ends, superficially tuberculate.

Ziziphus jujuba ssp. Jujuba: **Trees**; drupes globose or oblong, large (1.4–4 cm across), red to dark-red, shiny, epicarp thin, mesocarp thick and sweet, not forming a cavum when dry; stone large, oblongoid, acute at both ends, deeply tuberculate...

Local uses: Ripe fruits are edible. It is styptic, blood purifier and stomachic. Also used in the treatment of asthma, bronchitis, diabetes, eye diseases, inflammatory skin conditions, liver disorders, ulcers and wounds.

Propagation

Various methods are adopted to multiply this important medicinal fruit plant. Propagation by seed is one of the oldest methods where seeds before sowing are given different pre-treatments. One of the most common techniques of propagation is the use of suckers. Sucker is a plant part (shoot) growing from stem-originated root at the base of the tree, usually used for propagation and can preserve desirable characteristics such as growth rate, fruit yield, and flavor that is commonly used in jujube industry (Sharif *et al.*, 2015). Jujube contains high number of offshoots or suckers on lateral roots. Use of suckers is one of the most common techniques of jujube propagation (Johnstone, 2016). This technique is also being evaluated through research trials at faculty of forestry, SKUAST-K. Different sizes of suckers have been collected to assess the adoption of particular size foe mass multiplication. The preliminary results show that sucker size significantly affects the different growth parameters of suckers. Other methods of propagation include *in vitro* propagation, propagation by cutting, propagation by grafting etc.

References

Abbas, M. F. (1997). Jujube In: Post-harvest physiology and storage of Tropical and Sub tropical Fruits. (Mitra, K.S. ed.), CAB International, Wallingford, pp 405-415.

Akhter, C. Dar, G. H. and Khuroo, A. A. (2013). *Ziziphus jujuba* Mill. subsp. spinosa (Bunge) Peng, Li & Li: a New Plant Record for the Indian Subcontinent. *Taiwania*, **58** (2): 132–135.

Azam-Ali, S. Bonkoungou, E. Bowe, C. deCock, C. Godara, A. and Williams, J. T. (2006). Fruits for the Future 2: Ber and other Jujubes. International Centre for Underutilized Crops, University of Southampton, Southampton, UK.

Guo, Y. and Shan, G. (2010). The Chinese Jujube (Shanghai Scientific and Technical Publishers.

Hu, F. (2011). Study on Four Main Functional Components of Chinese Jujube. Hebei Agricultural University, Baoding.

Islam, M.B. and Simmons, M.P. (2006). A Thorny Dilemma: Testing Alternative Intrageneric Classifications within Ziziphus (Rhamnaceae). *Systematic Botany*. **31**: 826–842.

Johnstone, R. (2012). Growing Chinese jujubes in WA. http://www.agric.wa.gov.au/objtwr/imported_assets/content/hort/fn/cp/fn_growing_Chinese_jujubes_in_wa.pdf.

Johnstone, R. (2016). Propagating jujubes. Dept. Fd. Agr. Res. Library, 1-11.

Li, J. W. Ding, S. D. and Ding, X. L. (2005). Comparison of antioxidant capacities of extracts from five cultivars of Chinese jujube. *Process Biochemistry*. **40**: 3607-3613.

Liu, D., Ye, X. and Jiang, Y. (2016). Chinese Dates: A Traditional Functional Food (CRC press).

Liu, M. and Wang, J. (2019). Fruit scientific research in new China in the past 70 years: Chinese jujube. J. Fruit. Sci. 36, 1369–1381.

Liu, M. and Wang, M. (2009). Germplasm Resources of Chinese Jujube (China Forestry Publishing House).

Liu, M. J. Zhao, J. Cai, Q. L. Liu, G. C. Wang, J. R. Zhao, Z. H. Liu, P. Dai, L. Yan, G. Wang, W. J. Li, X. S. Chen, Y. Sun, Y. D. Liu, Z. G. Lin, M. J. Xiao, J. Chen, Y. Y. Li, X. F. Wu, B. Ma, Y. Jian, J. B. Yang, W. Yuan, Z. Sun, X. C. Wei, Y. L. Yu, L. L. Zhang, C. Liao, S. G. He, R. J. Guang, X. M. Wang, Z. Zhang Y. Y.and Luo, L. H. (2014). The complex jujube genome provides insights into fruit tree biology. *Nature Communications*. **5**: 1-13.

Liu, X., Li, H., Liu, Z., Zhao, J. and Liu, M. (2016). Genetic diversity and structure of 255 cultivars of *Ziziphus jujuba* Mill. Sci. Agric. Sin. 49, 2772–2791.

Sharif, N. Gill, J. I. Abbas, M. M. Javaid, M. A. and Memon, N. U. N. (2015). Effective propagation technique and time of grafting/budding in ber (*Ziziphus mauritiana* Lamk.). *Journal of agricultural research*. **53**: 83-92.

Stanica, F. (2019). Twenty years of jujube (*Ziziphus jujuba* Mill.) research in Romania. Sci. Pap. Ser. B Hortic. LXIII, 17–24.

Tahergorabi, Z., Abedini, M. R. Mitra, M. Fard M. H. and Beydokhti, H. (2015). *Ziziphus jujuba*: A red fruit with promising anticancer activities. *Pharmacognosy Reviews*. **9**: 99-106.

Xinguang, W. Bo, L. Chengjiu, G. Youke, W. Jianqiang, H. Liu, S. Tieliang, W. and Mingze, Y. 2017. Identification of sap flow driving factors of jujube plantation in semi-arid areas in Northwest China. *International Journal of Agricultural and Biological Engineering*. **10**: 172-183.